Various Applications of Methods and Elements of Adaptive Optics

Various Applications of Methods and Elements of Adaptive Optics

Editor

Julia Sheldakova

MDPI • Basel • Beijing • Wuhan • Barcelona • Belgrade • Manchester • Tokyo • Cluj • Tianjin

Editor
Julia Sheldakova
Atmospheric Adaptive Optics Lab
Institute of Geosphere
Dynamics RAS
Moscow
Russia

Editorial Office
MDPI
St. Alban-Anlage 66
4052 Basel, Switzerland

This is a reprint of articles from the Special Issue published online in the open access journal *Photonics* (ISSN 2304-6732) (available at: www.mdpi.com/journal/photonics/special_issues/adaptive_optics).

For citation purposes, cite each article independently as indicated on the article page online and as indicated below:

LastName, A.A.; LastName, B.B.; LastName, C.C. Article Title. *Journal Name* **Year**, *Volume Number*, Page Range.

ISBN 978-3-0365-6206-3 (Hbk)
ISBN 978-3-0365-6205-6 (PDF)

© 2023 by the authors. Articles in this book are Open Access and distributed under the Creative Commons Attribution (CC BY) license, which allows users to download, copy and build upon published articles, as long as the author and publisher are properly credited, which ensures maximum dissemination and a wider impact of our publications.

The book as a whole is distributed by MDPI under the terms and conditions of the Creative Commons license CC BY-NC-ND.

Contents

About the Editor vii

Preface to "Various Applications of Methods and Elements of Adaptive Optics" ix

Freddie Santiago, Carlos O. Font, Sergio R. Restaino, Syed N. Qadri and Brett E. Bagwell
Design, Fabrication and Characterization of an Adaptive Retroreflector (AR)
Reprinted from: *Photonics* **2022**, *9*, 124, doi:10.3390/photonics9030124 1

Qi Hu, Yuyao Xiao, Jiahe Cui, Raphaël Turcotte and Martin J. Booth
The Lattice Geometry of Walsh-Function-Based Adaptive Optics
Reprinted from: *Photonics* **2022**, *9*, 547, doi:10.3390/photonics9080547 11

Alexey Rukosuev, Alexander Nikitin, Vladimir Toporovsky, Julia Sheldakova and Alexis Kudryashov
Real-Time Correction of a Laser Beam Wavefront Distorted by an Artificial Turbulent Heated Airflow
Reprinted from: *Photonics* **2022**, *9*, 351, doi:10.3390/photonics9050351 29

Maksym Shpakovych, Geoffrey Maulion, Alexandre Boju, Paul Armand, Alain Barthélémy and Agnès Desfarges-Berthelemot et al.
On-Demand Phase Control of a 7-Fiber Amplifiers Array with Neural Network and Quasi-Reinforcement Learning
Reprinted from: *Photonics* **2022**, *9*, 243, doi:10.3390/photonics9040243 43

Pavel A. Khorin, Alexey P. Porfirev and Svetlana N. Khonina
Adaptive Detection of Wave Aberrations Based on the Multichannel Filter
Reprinted from: *Photonics* **2022**, *9*, 204, doi:10.3390/photonics9030204 55

Ilya Galaktionov, Alexander Nikitin, Julia Sheldakova, Vladimir Toporovsky and Alexis Kudryashov
Focusing of a Laser Beam Passed through a Moderately Scattering Medium Using Phase-Only Spatial Light Modulator
Reprinted from: *Photonics* **2022**, *9*, 296, doi:10.3390/photonics9050296 75

Yamin Zheng, Ming Lei, Shibing Lin, Deen Wang, Qiao Xue and Lei Huang
Filtered Influence Function of Deformable Mirror for Wavefront Correction in Laser Systems
Reprinted from: *Photonics* **2021**, *8*, 410, doi:10.3390/photonics8100410 85

Vladimir Toporovsky, Alexis Kudryashov, Arkadiy Skvortsov, Alexey Rukosuev, Vadim Samarkin and Ilya Galaktionov
State-of-the-Art Technologies in Piezoelectric Deformable Mirror Design
Reprinted from: *Photonics* **2022**, *9*, 321, doi:10.3390/photonics9050321 101

Feodor Kanev, Nailya Makenova and Igor Veretekhin
Development of Singular Points in a Beam Passed Phase Screen Simulating Atmospheric Turbulence and Precision of Such a Screen Approximation by Zernike Polynomials
Reprinted from: *Photonics* **2022**, *9*, 285, doi:10.3390/photonics9050285 111

Valentina Klochkova, Julia Sheldakova, Ilya Galaktionov, Alexander Nikitin, Alexis Kudryashov and Vadim Belousov et al.
Local Correction of the Light Position Implemented on an FPGA Platform for a 6 Meter Telescope
Reprinted from: *Photonics* **2022**, *9*, 322, doi:10.3390/photonics9050322 125

About the Editor

Julia Sheldakova

Dr. Julia Sheldakova is a leading researcher at IDG RAS, successfully working in the field of modern adaptive optics for 20 years. Her main scientific results are related to the development of methods of adaptive mirrors control to form a given intensity distribution of high-power laser radiation.

Preface to "Various Applications of Methods and Elements of Adaptive Optics"

Advances in the field of adaptive optics have produced an expanding toolkit for a growing number of photonics applications, including laser beam propagation, signal processing, vision science, astronomy, and other areas. Innovation in laser adaptive optics is key to solving various scientific and technological problems, from improving the performance of laser systems to enabling new applications. This volume is focused on a wide range of topics, including but not limited to the following:

- adaptive optic components and tools;
- wavefront sensing; control algorithms;
- beam shaping and control;
- imaging;
- astronomy;
- optical communications;
- propagation through turbulent and turbid media.

I wish to thank all the people who helped me with gathering papers and reviewing. The support of my colleagues is also greatly appreciated.

Julia Sheldakova
Editor

Communication

Design, Fabrication and Characterization of an Adaptive Retroreflector (AR)

Freddie Santiago [1,*], Carlos O. Font [1], Sergio R. Restaino [1], Syed N. Qadri [1] and Brett E. Bagwell [2]

1. Naval Research Laboratory, Washington, DC 20375, USA; carlos.font@nrl.navy.mil (C.O.F.); sergio.restaino@nrl.navy.mil (S.R.R.); noor.qadri@nrl.navy.mil (S.N.Q.)
2. Sandia National Laboratories, Albuquerque, NM 94551, USA; bbagwel@sandia.gov
* Correspondence: freddie.santiago@nrl.navy.mil

Abstract: Recent work at the U.S. Naval Research Laboratory studied atmospheric turbulence on dynamic links with the goal of developing an optical anemometer and turbulence characterization system for unmanned aerial vehicle (UAV) applications. Providing information on the degree of atmospheric turbulence, as well as wind information and scintillation, in a low size, weight and power (SWaP) system is key for the design of a system that is also capable of adapting quickly to changes in atmospheric conditions. The envisioned system consists of a bi-static dynamic link between a transmitter (Tx) and a receiver (Rx), relying on a small UAV. In a dynamic link, the propagation distance between the Tx/Rx changes rapidly. Due to SWaP constraints, a monostatic system is challenging for such configurations, so we explored a system in which the Tx/Rx is co-located on a mobile platform (UAV), which has a mounted retroreflector. Beam divergence control is key in such a system, both for finding the UAV (increased beam divergence at the Tx) and for signal optimization at the Rx. This led us to the concept of using adaptive/active elements to control the divergence at the Tx but also to the implementation of an adaptive/active retroreflector in which the return beam divergence can be controlled in order to optimize the signal at the Rx. This paper presents the design, fabrication and characterization of a low SWaP adaptive retroreflector.

Keywords: adaptive retroreflector; tunable lens; adaptive lens; polymer optics; divergence control; fluidic lens; tunable optics

Citation: Santiago, F.; Font, C.O.; Restaino, S.R.; Qadri, S.N.; Bagwell, B.E. Design, Fabrication and Characterization of an Adaptive Retroreflector (AR). *Photonics* **2022**, 9, 124. https://doi.org/10.3390/photonics9030124

Received: 25 January 2022
Accepted: 15 February 2022
Published: 22 February 2022

Publisher's Note: MDPI stays neutral with regard to jurisdictional claims in published maps and institutional affiliations.

Copyright: © 2022 by the authors. Licensee MDPI, Basel, Switzerland. This article is an open access article distributed under the terms and conditions of the Creative Commons Attribution (CC BY) license (https://creativecommons.org/licenses/by/4.0/).

1. Introduction

Retroreflectors are passive devices that return the incident signal through the same propagation path. For our intended application on UAVs, a retroreflector is ideal, due to its size and zero power consumption. The fact that this is a dynamic link (with changing distance between the transmitter and the point where the signal is reflected occurring quickly or discretely) means that signal degradation is expected due to atmospheric turbulence induced effects, but also due to the general nature of a propagating beam. In order to ameliorate these effects, we relied on low order adaptive optics correction, in this case, focus control. Due to the constraints in SWaP, we have designed and fabricated an adaptive retroreflector which allows us to change the divergence of the beam in order to optimize the link, achieving higher link performance or longer distances than can normally be obtained with a passive system. This device enables the control of the divergence, which can be used to optimize the return signal in a monostatic configuration or to increase the return footprint of the beam in a bi-static, dynamic, or reconfigurable link (moving link), this latter case being the motivation for the following types of devices [1,2].

Adaptive optical devices (also known as active or tunable devices) are devices that can adjust their surface/curvature (such as deformable mirrors, fluidic lenses, elastic/elastomeric solids) or modify their index of refractions (such as liquid crystals) in order to change the optical properties of the element, such as its focal length. This leads to the design and fabrication of an adaptive retroreflector—the one described in this paper is based on fluidic optical elements, for which NRL has extensive expertise [3–5].

2. AR Design and Configurations

Our adaptive retroreflector (AR) can be used with a corner cube retroreflector, solid or hollow, and consists of an optical fluid, encapsulated by an elastomeric membrane that can be deformed via an actuator—in this case, the same actuator we use for our adaptive polymer lenses. This actuator and its electronics have been designed for tactical applications in which SWaP is key, making this ideal for UAV applications.

Depending on the application, both the membrane and fluid can be replaced with an elastomeric optical polymer (which can be made from the same material as the optical membrane) that can be deformed mechanically to make the adaptive retroreflector. This device enables contro-l of the divergence.

The device can be fabricated in two ways: (1) using an elastomeric optical polymer, or (2) a fluidic adaptive/tunable device with a hollow or solid retroreflector. While manufacturing errors could change the operation of a retroreflector with an elastic polymer in front, by slightly changing the direction of return light, such errors can be easily quantified and corrected, for example, by monitoring the overall optical performance with an interferometer.

For the elastomeric optical polymer option, the elastic polymer is molded to a desired initial shape and the change on the polymer surface can be affected by means of applying a pressure/compression to the polymer. An elastomeric optical polymer is a polymer that has high transmission at the user-desired operational wavelength and has elastic properties which allow the solid substrate to be deformed. A second alternative to deform the polymer can be achieved by the use of dielectric elastomer actuation in which a voltage is applied to a pliable electrode and the polymer is deformed, creating the change on its surface. Figure 1 shows the configuration of the elastomer optical polymer, as well as three operational states of the adaptive retroreflector.

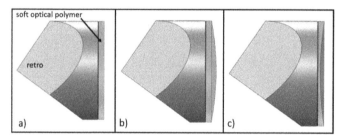

Figure 1. Schematic of an adaptive retroreflector using a solid elastomeric polymer: (**a**) flat (**b**) convex and (**c**) concave.

For a fluidic adaptive/tunable option, an elastomeric membrane encapsulates an optical fluid which is mounted on the front of a hollow or solid retroreflector. The elastomeric membrane needs to have similar optical and mechanical properties to those described above for the solid option. The optical fluid, needs to be optically and chemically compatible with the membrane and needs to have high transmission at the operational wavelength. Polydimethylsiloxane is a common polymer that can be used for the membrane as well as for the elastomeric solid option. For the optical fluids, there are numerous oils, polymers and resins that have been studied (for example, water, glycerol, etc.) [3]. The actuation of this system can be achieved by compressing/decompressing the flexible membrane, which creates a change on its surface. This occurs by moving a cylinder along the optical axis of the system, thus compressing the circumference of the flexible membrane. Besides the mechanical action, magnetic actuation or use of a compliant electrode (dielectric elastomer) can achieve actuation of the membrane. There are other actuation techniques that are situatable and could be implemented as well, such as those used by commercially available fluidic lenses, for example, Varioptics, Optotune and Holochip [6–12]. Figure 2, shows a conceptual sketch of the adaptive retroreflector based on the flexible membrane/fluidic concept.

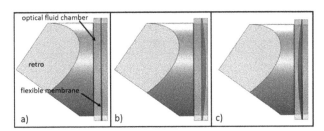

Figure 2. Schematic of an adaptive retroreflector using an optical liquid/fluidic: (**a**) flat (**b**) convex and (**c**) concave.

The latter configuration was selected for this paper. The device changes the divergence of the returned beam but can also work as a regular passive retroreflector if the system requires.

3. Fabrication, Characterization and Results

Here, we present the optical design of the AR as fabricated and in the configuration we used for testing. We describe the optical setups used for testing and comparing the device to a passive retroreflector. Data for the repeatability and arbitrary radius of curvature measurements was acquired using a Zygo Verifire HD optical interferometer. Furthermore, we show images taken comparing a passive retroreflector in comparison with AR, both actuated and flat.

3.1. AR Optical Design

We used OpticStudio nonsequential tools to model the corner cube retroreflector and adaptive components and, for visualization purposes, a beam splitter cube was added, as shown in Figure 3. This was also the configuration chosen for the test. Note that we did not model the thickness of the membrane, since the effects of the membrane are negligible in OpticStudio for this type of application. The model was performed using the volume of the fluid, the fluid acting as a lens which changes its radius of curvature and center thickness. The figure shows a collimated beam, incident on a beam splitter cube which reflects part of the incident light and transmits a portion, which then impinges on the adaptive retroreflector. Light is reflected back from the retroreflector and reflected again from the beam splitter cube and incident on the detector. In field operations, the beam splitter can be used to monitor the incoming beam and direct the adaptive retroreflector in order to control the divergence. It can also be used without the beam splitter cube, such that the beam can be monitored at the receiver side and the system optimized in a power-in-the-bucket (PIB) configuration using well-known algorithms (e.g., stochastic parallel-gradient-descent) [13,14]; the AR is then instructed by this information.

3.2. Fabrication of the AR

For this particular design we used a 12.7 mm corner cube retroreflector from Thorlabs, polydimethysiloxane (PDMS) as the membrane, glass support structures for the PDMS, and an optical fluid with an index of refraction of 1.45 and an Abbe number of 45.0 (at λ = 589 nm). The first step consisted in making the PDMS membrane which was then bonded to the glass support structure. The fluid was added to the membrane/glass structure and the corner cube was bonded to it. The last step was to mount the AR into the actuator and start the testing—the assembly steps are shown in Figure 4. An important note: for this proof of concept, we did not follow the special fabrication procedures that we normally utilize to reduce the surface wavefront error which involve the reduction of coma induced by gravity and astigmatism resulting from the materials and fabrication procedures. Figure 4 shows the top-level schematic of the assembly procedure, as well as pictures of the assembled AR (in its actuator). The actuator was custom-made for our adaptive lenses, and we were able to modify one to accommodate the AR. The actuator consists of a modified motor in a custom housing, with a maximum clear aperture of 19.5 mm, optical encoder with a resolution of ~50 nm, speed of ~2.5 mm/s, peak power consumption ~15 W, idle power consumption of ~1–500 µW and temperature monitoring of 0.01 °C. The electronics can control two actuators at the same time and can run off three CR-123 batteries (two batteries for a single actuator) with an average number of actuations of about 6000 per set of batteries.

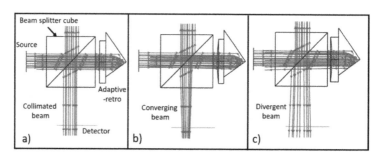

Figure 3. Optical design setup used for testing including the collimated source incident on the beam splitter cube with the outputs: (**a**) a collimated beam, (**b**) a converging beam, and (**c**) a divergent beam.

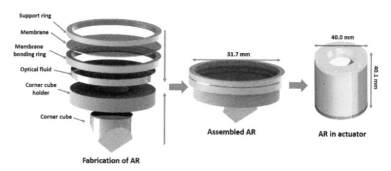

Figure 4. Schematic representation of the fabrication and assembly process of the AR.

3.3. Laboratory Optical Setup

The AR was tested in various ways. Firstly, in the same way that we measured the radius of curvature (ROC) of lenses using a Zygo interferometer—for this device we measured two ROCs as a test, a positive (convex surface) with a ROC of 234 mm and a negative (concave surface) of −185 mm. The second setup was to compare the performance and proof of concept of the AR in comparison with a passive retroreflector, as shown in Figure 5. We used the HeNe 633 nm source of the Zygo interferometer and a 1550 nm was co-aligned for further testing. Data from the 1550 nm was not included but performance of the active surface component at this wavelength has been demonstrated in a previous report [5]. We were able to use the beam collimated or with the addition of a known divergence that could be removed with the AR and compared with the passive retroreflector. The setup with the beam splitter cube allowed us to look at the return beam with the interferometer and, on the other arm, to look at the output with a camera, photodetector or power meter, while we were able to use beam blocks to look at each retroreflector individually or additionally, enabling viewing of the interference fringes formed by the two. This facilitated alignment, but also monitoring of the difference when the AR is actuated. This same setup was also used to perform first order measurements of the repeatability of the AR by actuating the AR from a flat state to a convex or concave state and back to a flat state, while measuring the surface form with the interferometer. An important note: temperature was monitored in the room, but temperature compensation of the AR was not used—the room environment was stable and thus compensation was not required.

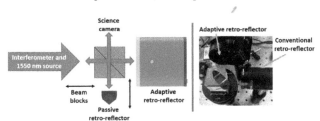

Figure 5. (**Left**) Layout of the optical testing setup. (**Right**) Picture of optical devices used for measurements.

3.4. Results/Discussion

3.4.1. ROC and Surface Measurements

The first test consisted of measurements of the ROC, positive and negative, in order to evaluate the performance of the device. Figure 6 shows a set of measurements, including a (left) measurement for a positive ROC of 255 mm and a (right) measurement for a negative ROC of 184 mm. The top row shows the 3D surface profile, and the bottom row indicates the 2D profile. The black circles in the figure are software masks used to remove unnecessary back reflections created by dust particles in the reference sphere. Within the respective figures, the left column (A or C) is the raw measurement and the right side (B or D) is with the dominant aberrations removed. As mentioned before, fabrication was not optimized for the surface figure, but what can be seen is the typical dominant aberration of coma and astigmatism, which are characteristic for this type of fluidic structures. Coma is due to gravity and astigmatism is due to fabrication or assembly procedures. For the positive ROC case demonstrated below, coma is the dominant aberration. On the negative ROC, there is a combination of coma and astigmatism, because measurements were taken close to the negative resting ROC (fabricated ROC) of the membrane for the fabricated AR device. The fabricated aberrations were more noticeable closer to the resting ROC because, for this type of actuation mechanism, this is the point of contact where boundary conditions are established between the membrane and actuation surface for the clear aperture. At this point, the amplitude of any existing aberrations can be enhanced. Another aberration that can be noticed is trefoil on both ROCs—this was purely due to the assembly in the actuator. We developed procedures for fabrication and assembly that reduce the dominant aberrations which are implemented when building adaptive polymer lenses, with the caveat that we can minimize coma based on the application, but do not completely eliminate it. The procedure to eliminate coma during fabrication is extremely complex, costly and time consuming if performed at the active surface. There are other ways to minimize it, including using a corrective element along the optical path of the system or close to the active surface, and this is a typical configuration used in commercial adaptive/tunable lenses [11].

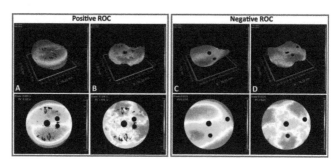

Figure 6. (**Left**) Positive ROC measurement (**A,B** columns) 2D and 3D profiles with dominant aberration removed, in this instance being coma (**B** column). (**Right**) Negative ROC measurement (**C,D** columns) 2D and 3D profiles with dominant aberration removed, in this instance, coma and astigmatism (**D** column).

Using the data from Figure 6, a Zernike fit was performed using the Mx software tools from the interferometer and coefficients of the fit for both the positive and negative ROCs are shown in Table 1.

Table 1. Results from the Zernike fit coefficients obtained from the data in Figure 6, for positive ROC (top) and negative ROC (bottom).

Coeff	Value (λ)	n	m	Representation
ROC = 225 mm				
Zernike Fit				
ZFR 0	0.000	0	0	1
ZFR 1	0.000	1	1	$\rho\cos(\theta)$
ZFR 2	0.000	1	−1	$\rho\sin(\theta)$
ZFR 3	0.018	2	0	$-1 + 2\rho^2$
ZFR 4	−0.085	2	2	$\rho^2\cos(2\theta)$
ZFR 5	−0.407	2	−2	$\rho^2\sin(2\theta)$
ZFR 6	−0.122	3	1	$(-2\rho + 3\rho^3)\cos(\theta)$
ZFR 7	2.191	3	−1	$(-2\rho + 3\rho^3)\sin(\theta)$
ZFR 8	−0.047	4	0	$1 - 6\rho^2 + 6\rho^4$
ROC = −184 mm				
Zernike Fit				
ZFR 0	0.000	0	0	1
ZFR 1	0.000	1	1	$\rho\cos(\theta)$
ZFR 2	0.000	1	−1	$\rho\sin(\theta)$
ZFR 3	−0.049	2	0	$-1 + 2\rho^2$
ZFR 4	−0.211	2	2	$\rho^2\cos(2\theta)$
ZFR 5	−1.823	2	−2	$\rho^2\sin(2\theta)$
ZFR 6	−0.108	3	1	$(-2\rho + 3\rho^3)\cos(\theta)$
ZFR 7	−2.115	3	−1	$(-2\rho + 3\rho^3)\sin(\theta)$
ZFR 8	−0.372	4	0	$1 - 6\rho^2 + 6\rho^4$

Figure 7, shows data taken for the AR at the same ROCs mentioned above but in a perpendicular configuration in order to eliminate the effects of coma due to gravity. Note, that for the data no terms have been removed. Astigmatism and trefoil were noticeable but the large magnitude due to coma was absent.

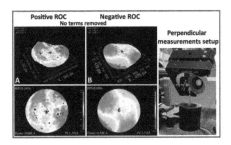

Figure 7. (**A**) Positive and (**B**) negative ROC, 2D and 3D surface representation for the perpendicular setup. Right side shows a picture of the setup.

The same procedure was performed on the results from Figure 7 and Table 2 shows the Zernike fit coefficients for the perpendicular measurements.

Table 2. Results from the Zernike fit coefficients obtained from the data in Figure 7, for positive ROC (top) and negative ROC (bottom).

Coeff	Value (λ)	n	m	Representation
ROC = 225 mm Perpendicular				
Zernike Fit				
ZFR 0	0.000	0	0	1
ZFR 1	0.000	1	1	$\rho\cos(\theta)$
ZFR 2	0.000	1	−1	$\rho\sin(\theta)$
ZFR 3	0.038	2	0	$-1 + 2\rho^2$
ZFR 4	0.137	2	2	$\rho^2\cos(2\theta)$
ZFR 5	−0.390	2	−2	$\rho^2\sin(2\theta)$
ZFR 6	0.065	3	1	$(-2\rho + 3\rho^3)\cos(\theta)$
ZFR 7	−0.071	3	−1	$(-2\rho + 3\rho^3)\sin(\theta)$
ZFR 8	−0.099	4	0	$1 - 6\rho^2 + 6\rho^4$
ROC = −184 mm Perpendicular				
Zernike Fit				
ZFR 0	0.000	0	0	1
ZFR 1	0.000	1	1	$\rho\cos(\theta)$
ZFR 2	0.000	1	−1	$\rho\sin(\theta)$
ZFR 3	−0.202	2	0	$-1 + 2\rho^2$
ZFR 4	−0.052	2	2	$\rho^2\cos(2\theta)$
ZFR 5	−1.248	2	−2	$\rho^2\sin(2\theta)$
ZFR 6	0.126	3	1	$(-2\rho + 3\rho^3)\cos(\theta)$
ZFR 7	−0.112	3	−1	$(-2\rho + 3\rho^3)\sin(\theta)$
ZFR 8	−0.224	4	0	$1 - 6\rho^2 + 6\rho^4$

3.4.2. Repeatability Measurements

Repeatability measurements were taken using the setup in Figure 5. The data collection consisted in changing the actuation state by a known encoder count from a flat state to a convex/concave state, while recording the encoder position as well as the data from the inferferometer. The encoder data is in the form of a set of three numbers: the set position by user (state of the lens), the temperature compensate position (once thermal compensation is activated) and the measured position. This last position, or the difference from the set position, was recorded. Readings from the interferometer PV(λ) (peak to valley) and power(λ) were recorded as well. Data was taken for a delta for encoder counts of 150 and 300 from flat, on both positive (convex or higher encoder counts), and negative (concave or lower encoder counts) direction. Figure 8 shows a sequence of consecutive measurements from flat to positive and Figure 9 shows measurements from flat to negative for a delta of 150 encoder counts. Note, the hexagonal pattern was a result of the facets of the corner cube. This was noticeable in this configuration based on the testing setup with the interferometer using a transmission flat. For the ROCs the measurements differed, since we were using a reference sphere and the spherical wavefront matched the deformed membrane, not the retroreflector.

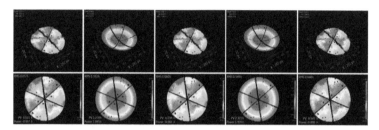

Figure 8. The AR was actuated from the flat state to a compressed state (or convex surface) and back to flat. For each case the top row is the 3D surface and the bottom row the 2D surface.

Figure 9. The AR was actuated from the flat state to a less compressed state (or concave surface) and back to flat. For each case the top row is the 3D surface and the bottom row the 2D surface.

In Figure 10, data is presented in graphical (with error bars based on the standard deviation), and tabular, form for the sequences, with 20 data points for delta 150 and 10 points for delta 300, and all cases starting from the same initial flat position. The average and standard deviation for the encoder position and peak-to-valley for the cases are shown in the table.

Figure 11 shows a comparison of the AR with a passive retroreflector. A screen was placed at a distance of approximately 1500 mm and the response from a collimated beam recorded and the AR was actuated in order to focus the beam on the screen.

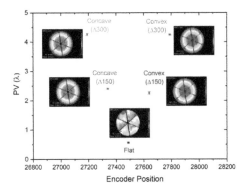

	Flat	PV(λ)	Convex (Δ=150)	PV(λ)	Convex (Δ=300)	PV(λ)	Concave (Δ=150)	PV(λ)	Concave (Δ=300)	PV(λ)
Ave. encod. Pos	27,490	0.564	27,640	2.261	27,790	4.248	27,340	2.408	27,190	4.256
Stdev	0.16	0.02	0.00	0.05	0.00	0.02	0.00	0.02	0.00	0.04

Figure 10. Graphical and tabular representation of the actuation sequence for the cases described above. Delta values refer to the change in encoder position from the flat state.

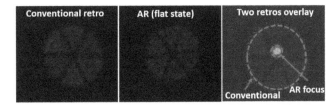

Figure 11. Images taken in the laboratory at a known distance from the retroreflector are shown (**left**) conventional retroreflector, (**middle**) AR in the flat state, and (**right**) both retroreflectors overlapping in the screen with the AR actuated to focus the beam at that particular distance.

4. Conclusions

We have presented the concept of an adaptive retroreflector. This concept was developed during a data campaign to study the atmospheric turbulence in a dynamic link, with the end goal of an optical anemometer for UAV applications in which the propagation distance is changing rapidly. The concept of the AR was then designed, fabricated and tested in a laboratory environment as a proof of concept. This particular device can operate from the VIS to the SWIR and preliminary parameters of its performance were studied. The next step will consist of fabricating a device following the tighter tolerance procedures developed previously for adaptive lenses. A follow-up report will consist of performing a calibration in a laboratory environment, including thermal compensation and quantification of losses added by absorption and/or scattering due to the membrane/fluid combination in comparison with a conventional retroreflector. The latter case will be studied in more detail in a field experiment where we can compare the losses due to the addition of the membrane/fluid combination with the losses of a conventional retroreflector (e.g., due to diffraction, or divergence introduced by atmospheric turbulence) at a propagation path. A power-in-the-bucket configuration will be used to compare the divergence control of the adaptive retroreflector and a conventional one. While the overall losses depend on configuration and materials, our experience with fluidic lenses has shown that the transmission losses are negligible compared to effects induced by turbulence.

5. Patents

A provisional patent application has been submitted, U.S. Patent Application Serial No. 62/695,310.

Author Contributions: Conceptualization, F.S. and C.O.F.; methodology, F.S., C.O.F., B.E.B. and S.R.R.; formal analysis, F.S., S.N.Q. and S.R.R.; writing—original draft preparation, F.S., C.O.F., S.R.R.; writing—review and editing, S.R.R., S.N.Q. and B.E.B.; funding acquisition, F.S. All authors have read and agreed to the published version of the manuscript.

Funding: This research received no external funding.

Institutional Review Board Statement: Not applicable.

Informed Consent Statement: Not applicable.

Data Availability Statement: Data is contained within the article.

Conflicts of Interest: The authors declare no conflict of interest.

References

1. Font, C.; Apker, T.; Santiago, F. Laser Anemometer for Autonomous Systems Operations. In *AIAA Infotech@ Aerospace*; AIAA SciTech: San Diego, CA, USA, 2016; p. 1230. [CrossRef]
2. Font, C.; Santiago, F.; Apker, T. Atmospheric Turbulence Measurements in Dynamical Links. In *Propagation Through and Characterization of Atmospheric and Oceanic Phenomena, OSA Technical Digest (Online)*, 3rd ed.; Paper M2A.3.; Optical Society of America: Washington, DC, USA, 2016; pp. 154–196.
3. Santiago, F.; Bagwell, B.; Martinez, T.; Restaino, S.; Krishna, S. Large aperture adaptive doublet polymer lens for imaging applications. *J. Opt. Soc. Am. A* **2014**, *31*, 1842–1846. [CrossRef]
4. Santiago, F.; Font, C.; Restaino, S. Adaptive Polymer Lenses at NRL. In *Applied Industrial Optics 2019, OSA Technical Digest*; Paper T2A.2.; Optical Society of America: Washington, DC, USA, 2019.
5. Santiago, F.; Bagwell, B.E.; Pinon, V., III; Krishna, S. Adaptive polymer lens for rapid zoom shortwave infrared imaging applications. *Opt. Eng.* **2014**, *53*, 125101. [CrossRef]
6. Berge, B.; Peseux, J. Variable focal lens controlled by an external voltage: An application of electrowetting. *Eur. Phys. J. E* **2000**, *3*, 159–163. [CrossRef]
7. Berge, B. Electrocapillarite et mouillage de films isolants par l'eau. *C.R. Acad. Sci. Ser. II Mec. Phys. Chim. Sci. Terre Univ.* **1993**, *317*, 157.
8. Mugele, F.; Baret, J. Electrowetting: From basics to applications. *J. Phys. Condens. Matter IOP Publ.* **2005**, *17*, R705–R774. [CrossRef]
9. Heikenfeld, J.; Smith, N.; Dhindsa, M.; Zhou, K.; Kilaru, M.; Hou, L.; Zhang, J.; Kreit, E.; Raj, B. Recent Progress in Arrayed Electrowetting Optics. *Opt. Photonics News* **2009**, *20*, 20–26. [CrossRef]
10. Yiu, J.; Batchko, R.; Robinson, S.; Szilagyi, A. A fluidic lens with reduced optical aberration. In *Intelligent Robots and Computer Vision XXIX: Algorithms and Techniques*; Proc. SPIE 8301; SPIE: Burlingame, CA, USA, 2012; p. 830117.
11. Blum, M.; Büeler, M.; Grätzel, C.; Giger, J.; Aschwanden, M. Optotune focus tunable lenses and laser speckle reduction based on electroactive polymers. In *MOEMS and Miniaturized Systems XI*; Proc. SPIE 8252; SPIE: San Francisco, CA, USA, 2012; p. 825207.
12. Vorontsov, M.A.; Sivokon, V.P. Stochastic parallel-gradient-descent technique for high-resolution wave-front phase-distortion correction. *J. Opt. Soc. Am. A* **1998**, *15*, 2745–2758. [CrossRef]
13. Font, C.O.; Gilbreath, G.C.; Bajramaj, B.; Kim, D.S.; Santiago, F.; Martinez, T.; Restaino, S.R. Characterization and training of a 19-element piezoelectric deformable mirror for lensing. *J. Opt. Fiber. Commun. Res.* **2010**, *7*, 1–9. [CrossRef]
14. Ma, S.; Yang, P.; Lai, B.; Su, C.; Zhao, W.; Yang, K.; Jin, R.; Cheng, T.; Xu, B. Adaptive Gradient Estimation Stochastic Parallel Gradient Descent Algorithm for Laser Beam Cleanup. *Photonics* **2021**, *8*, 165. [CrossRef]

Article

The Lattice Geometry of Walsh-Function-Based Adaptive Optics

Qi Hu, Yuyao Xiao, Jiahe Cui, Raphaël Turcotte and Martin J. Booth *

Department of Engineering Science, University of Oxford, Oxford OX1 3PJ, UK
* Correspondence: martin.booth@eng.ox.ac.uk

Abstract: We show that there is an intrinsic link between the use of Walsh aberration modes in adaptive optics (AO) and the mathematics of lattices. The discrete and binary nature of these modes means that there are infinite combinations of Walsh mode coefficients that can optimally correct the same aberration. Finding such a correction is hence a poorly conditioned optimisation problem that can be difficult to solve. This can be mitigated by confining the AO correction space defined in Walsh mode coefficients to the fundamental Voronoi cell of a lattice. By restricting the correction space in this way, one can ensure there is only one set of Walsh coefficients that corresponds to the optimum correction aberration. This property is used to enable the design of efficient estimation algorithms to solve the inverse problem of finding correction aberrations from a sequence of measurements in a wavefront sensorless AO system. The benefit of this approach is illustrated using a neural-network-based estimator.

Keywords: lattice geometry; Walsh functions; adaptive optics

1. Introduction

Many adaptive optics (AO) methods have been developed to compensate phase aberrations in a range of applications including astronomy, ophthalmology and microscopy [1–3]. All AO systems are limited, in some way, by the capabilities of the adaptive element, typically a deformable mirror (DM) or a spatial light modulator (SLM), that corrects the aberrations. One such limitation is in the range of phase functions that the element can correct. The correction space of an AO element is defined by the range of phase functions that can be imparted by the device. For a pixelated AO device, such as a SLM or segmented DM, the correction space is defined by the set of accessible pixel states, which could be represented by the set of phase values for each pixel.

In many AO systems, it is preferable to design the system around a set of orthogonal modes for representation and control of the wavefront, rather than localized wavefront modulations. For example, wavefont-sensorless AO systems usually use a modal basis [4,5]. This method involves the sequential application of predetermined bias aberrations, the acquisition of a set of measurements of an appropriate quality metric, and then estimation of the required correction aberration. The conventional approach to sensorless AO is to use knowledge of the forward problem—that is how the quality metric is affected by input aberrations—to inform the design of an efficient estimation scheme that, in effect, solves the inverse problem of finding the optimal correction aberrations from the set of metric measurements. Such estimation can be performed using optimisation algorithms or neural networks (NN) to solve the inverse problem [6,7].

It is known that control using modes defined across the whole pupil provides stronger modulation of the optimisation metric than individual pixels or subregions of the pupil [5]. Such whole-pupil modulation hence provides more robust operation, particularly in low-light level imaging scenarios when the signal-to-noise ratio (SNR) is low. For such pixel-based sensorless AO systems, Walsh modes are an appropriate choice. Walsh modes are a set of orthogonal functions that represent phase patterns across a pixelated pupil, where the number of pixels is equal to a power of 2 [8,9]. Each Walsh mode consists of an equal

Citation: Hu, Q.; Xiao, Y.; Cui, J.; Turcotte, R.; Booth, M.J. The Lattice Geometry of Walsh-Function-Based Adaptive Optics. *Photonics* **2022**, *9*, 547. https://doi.org/10.3390/photonics9080547

Received: 10 June 2022
Accepted: 30 July 2022
Published: 4 August 2022

Publisher's Note: MDPI stays neutral with regard to jurisdictional claims in published maps and institutional affiliations.

Copyright: © 2022 by the authors. Licensee MDPI, Basel, Switzerland. This article is an open access article distributed under the terms and conditions of the Creative Commons Attribution (CC BY) license (https://creativecommons.org/licenses/by/4.0/).

number of pixels taking each of the binary values $+1$ or -1. For the different modes, a different combination of pixels takes the positive and negative values.

The binary nature of Walsh modes means that the range of each mode that must be searched to find the optimum correction is finite. This contrasts with other modal bases built upon continuous functions, such as Zernike polynomials, which would have unbounded range (albeit limited by the stroke of the adaptive correction element).

However, care needs to be taken when considering combinations of Walsh modes, as multiple combinations of modes can have the same effect on the system. This means that there are multiple potential solutions to the inverse problem of finding the optimal set of Walsh mode coefficients that optimise aberration correction. These multiple solutions can cause complications in defining an estimator to solve the inverse problem. Solving the inverse problem would be considerably simplified if we could ensure that there was only one optimal solution in the search space.

We show that there are properties of the Walsh modes that link the operation of these sensorless AO systems to the mathematics of lattices [10]. We discuss how these mathematical properties can aid the design of aberration estimation algorithms by constraining the search space. Specifically, we show heuristically that through understanding of the lattice geometry, we can define a unique finite search space, in terms of combinations of Walsh mode coefficients, that contains a single optimum correction. This permits the implementation of an efficient NN-based optimisation scheme that can measure and correct any combination of N Walsh modes of any coefficient value using only $2N+1$ metric measurements. We show that a simple NN can be trained to solve the inverse problem if the search space is constrained using the lattice model, whereas the correct combination of Walsh mode coefficients cannot reliably be found for a nonconstrained search space.

2. Optical System Model

For the purposes of modelling, we considered the simple sensorless adaptive optics system shown in Figure 1. Such a model has been extensively used for analysis of such sensorless systems [11,12], as the principle of operation is readily extendable to similar optical systems, including applications in laser material processing, free-space communications, and laser scanning microscopy.

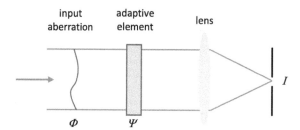

Figure 1. Optical system used for modelling. The input wavefront contains a phase aberration Φ, which passes through a correction device imparting an additional phase Ψ. The beam is focussed onto a vanishingly small pinhole detector on the optical axis that admits the intensity I.

The input beam to the system is collimated and has uniform amplitude. The input phase aberration is $\Phi(\mathbf{r})$ and the phase $\Psi(\mathbf{r})$ is added by the adaptive element (AE), which could be a pixelated SLM or a segmented DM. These are both considered to be added at the pupil P, which is taken to have unit radius; $\mathbf{r}$ is the normalised coordinate vector in the pupil. The lens performs a Fourier transform of the pupil field to provide a focal field. A vanishingly small pinhole detector is placed on axis at the centre of the focus and detects a signal I that corresponds to the on-axis intensity at the focus. This is equivalent to the power of the zero-frequency component of the Fourier transform, which is equivalent to the squared modulus of the mean value of the pupil field. Mathematically, this can be expressed as

$$I = \left| \frac{1}{\pi} \iint_P \exp i[\Phi(\mathbf{r}) + \Psi(\mathbf{r})] d^2\mathbf{r} \right|^2 \qquad (1)$$

where the maximum, aberration-free signal is normalised to 1.

3. Representation of Aberrations as Walsh Modes

For simplicity, let us assume $\Phi(\mathbf{r}) = 0$ so that all aberrations can be represented within $\Psi(\mathbf{r})$. Let us also assume that the adaptive element is a pixelated device, where each of the N pixels can introduce a piston phase. The phase introduced by the adaptive element could be expressed as

$$\Psi(\mathbf{r}) = \sum_{l=1}^{N} \alpha_l \eta_l(\mathbf{r}) \qquad (2)$$

where $\eta_l(\mathbf{r})$ are the phase influence functions of each pixel, which have value 1 within the pixel area and 0 elsewhere; α_l are the coefficients of these influence functions, which correspond to the pixel phase value. Alternatively, we could represent the AE phase as

$$\Psi(\mathbf{r}) = \sum_{k=0}^{N-1} \beta_k \omega_k(\mathbf{r}) \qquad (3)$$

where $\omega_k(\mathbf{r})$ are functions that take binary values of -1 or $+1$ in each pixel region, such that the lth pixel takes on the lth value of the kth Walsh function of length N, $W_k^N[l]$ [8]. β_k are the coefficients of these functions $\omega_k(\mathbf{r})$. For a given sequence length $N = 2^\gamma$, where γ is an integer, there are N orthogonal Walsh functions, each of which consists of $N/2$ elements of value -1 or $+1$, except for the first function that consists entirely of 1 s (see examples in Figure 2). We follow the convention that the Walsh function index starts at $k = 0$. From the above relationships, it is clear that each pixel phase can be represented as

$$\alpha_l = \sum_{k=0}^{N-1} \beta_k W_k^N[l] \qquad (4)$$

or alternatively in matrix–vector format as

$$\mathbf{a} = \mathbf{W}^T \mathbf{b} \qquad (5)$$

where **a** is a vector of length N that contains the phase value of each pixel, **b** is a vector of length N that contains the coefficient of each Walsh function and **W** is an $N \times N$ Walsh–Hadamard matrix consisting of values ± 1 [13,14]. The rows of this matrix correspond to each of the Walsh functions. The matrix **W** provides the mapping between the Walsh coefficients and the pixel values. For Hadamard matrices, $\mathbf{WW}^T = N\mathbf{I}_N$, where $\mathbf{I}_N$ is the identity matrix of size N [13,14]. Hence, we can invert Equation (5) as

$$\mathbf{b} = \frac{1}{N} \mathbf{W} \mathbf{a} \qquad (6)$$

Note that for a set of Walsh functions to be defined, we require $N = 2^\gamma$, where γ is an integer. We assume throughout this paper therefore that $N = 2^\gamma$. However, Hadamard matrices also exist for $N = 4\gamma$, where γ is an integer [13,14]. For simplicity, these other matrices will not be considered in this current analysis.

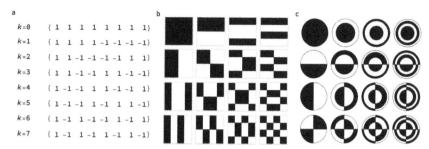

Figure 2. Illustrations of Walsh modes. (**a**) The Walsh functions $W_k^8[l]$ shown in numerical form. (**b**) The Walsh modes $W_k^{16}[l]$ shown as aberration basis modes over a square aperture. (**c**) The polar Walsh modes, equivalent to $W_k^{16}[l]$ shown as aberration basis modes over a circular aperture. In both (**b**) and (**c**), index $k = 0$ for the first mode is in the top left, and k increments in row-major order to $k = 15$ in the bottom right.

4. Lattice Symmetry of Aberration States

We can define the aberration state fully as the vector of pixel values **a**. Thus, we can consider that a point at position **a** in the N-dimensional space of pixel values is equivalent to an aberration state where any coefficient α_l is replaced by $\alpha_l + 2\pi q$, where q is an arbitrary integer. This reveals a repetitive structure in each coordinate of the N-dimensional space that results in a lattice type symmetry. Hence, in this space, there is an infinite number of points that represent a given aberration state and these points are arranged in a transformed integer lattice $\mathbb{Z}_N$ that is scaled by a factor of 2π and offset by the pixel value α_l along each dimension. Furthermore, the lattice structure is based around a fundamental unit that is an N-dimensional cube of side length 2π; this fundamental unit is known as a Voronoi cell [10].

The matrix–vector operation of Equation (6) can now be interpreted as a rotation (as **W** is an orthogonal matrix) and scaling by $1/N$ of the vector **a** to give the Walsh coefficient vector **b**. The lattice symmetry is hence maintained in a rotated and scaled form when the state is described by **b**. When represented by the vector **a**, any Walsh function consists of equal magnitude amounts ($+1$ or -1) of each pixel value, so the vector must be directed along certain body diagonals of the cubic Voronoi cell. After transformation, these body diagonals lie along the axes of the vector space containing **b**. This lattice symmetry will be used for derivations later in this article.

5. Effects of Pixels and Modes on Signal Modulation

Let us assume that each pixel of the AE has equal area (the pixels should have equal area if the amplitude profile at the pupil is uniform. For nonuniform illumination, the pixel area should be varied to provide the same total power in each pixel (e.g., the pixels could be large near the edge of the pupil for a Gaussian illumination profile.) No constraint is placed here on the position or shape of the pixels.), so that the integration used in Equation (1) can be replaced by a summation, assuming here that $\Phi(\mathbf{r}) = 0$:

$$I = \left| \frac{1}{N} \sum_{l=1}^{N} \exp(i\alpha_l) \right|^2 = \left| \frac{1}{N} \sum_{l=1}^{N} \exp\left(i \sum_{k=0}^{N-1} \beta_k W_k^N[l]\right) \right|^2 \qquad (7)$$

If the arbitrarily chosen lth pixel is varied and all other pixel values have the same value (here arbitrarily set to zero), then

$$I(\alpha_l) = \left(1 - \frac{D}{2}\right) + \frac{D}{2} \cos \alpha_l \qquad (8)$$

where the modulation depth $D = 4(N-1)/N^2$. For all other Walsh functions other than

$W_0^N[l]$, the signal is also cyclic as a function of β_k:

$$I(\beta_k) = \cos^2 \beta_k \qquad (9)$$

where the modulation depth has the maximum possible value of 1 and the period in terms of β_k is π. The effects of single pixel and modal variations are shown in Figure 3.

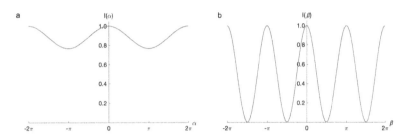

Figure 3. (a) Variation of signal for a 16-pixel system with a single pixel modulation showing low modulation depth and period 2π. (b) variation of a Walsh mode in the same system showing full modulation and period π.

If a combination of Walsh modes is present with small coefficients, we can use a Maclaurin expansion of the exponential in Equation (7) to give

$$I \approx 1 - \sum_j \sum_k \beta_j \beta_k \left(\frac{1}{N} \sum_l W_j^N[l] W_k^N[l] \right) + \left[\sum_k \beta_k \left(\frac{1}{N} \sum_l W_k^N[l] \right) \right]^2 \qquad (10)$$

The term in the final bracket $\frac{1}{N} \sum_l W_k^N[l]$ is equal to zero except for when $k = 0$, in which case it has value 1. The orthogonality property of the Walsh functions means that $\frac{1}{N} \sum_l W_j^N[l] W_k^N[l]$ in the second term is equivalent to the Kronecker delta function δ_{jk}. Hence, the signal is approximately

$$I \approx 1 - \sum_{k=1}^{N-1} \beta_k^2 \qquad (11)$$

This is equivalent to the well-known approximation of the Strehl ratio as $1 - \phi_{rms}^2$, where ϕ_{rms} is the root mean square value of the aberration, which in this case is equal to $\sqrt{\sum_k \beta_k^2}$.

6. Defining a Well-Corrected System

If we define our system to be "well-corrected" when the root-mean-square (rms) phase error is below a chosen value, such that $\phi_{rms} \leq \epsilon$, then the system will be well-corrected when $I \geq 1 - \epsilon^2$. We can also express the second term in Equation (11) as a length $N - 1$ vector $\mathbf{b}'$, which is equivalent to $\mathbf{b}$ with the piston coefficient removed, as

$$I \approx 1 - |\mathbf{b}'|^2 \qquad (12)$$

Our condition for being well-corrected is hence equivalent to requiring that $|\mathbf{b}'| \leq \epsilon$. Interpreted geometrically, this means that any point within an $(N-1)$-dimensional spherical volume of radius ϵ centred on the point where $I = 1$ will be considered well-corrected.

In practice, the total aberration in a system will be the sum of the input aberration and that introduced by the AE, that is $\Phi(\mathbf{r}) + \Psi(\mathbf{r})$. The values of $\mathbf{b}'$ discussed here consolidate these two sources of aberrations to represent the residual aberrations, such that we seek a perfect correction for which $\mathbf{b}' = 0$. We will also assume for this analysis that the input aberration consists entirely of modes that can be corrected by the AE.

7. Vector Space Representation of Well-Corrected States

The signal variation as a function of pixel values was given in Equation (7) and can be expressed alternatively as

$$I = \left| \frac{1}{N} \sum_{l=1}^{N} \exp(i\alpha_l) \right|^2 = \frac{1}{N^2} \left| 1 + \sum_{l=2}^{N} \exp(i\alpha'_l) \right|^2 \quad (13)$$

where $\alpha'_l = \alpha_l - \alpha_1$. From this, we find the value $I = 1$ can only be obtained if $\alpha'_l \mod 2\pi = 0$ for all l. We derive this result by considering the phasor sum of each of the terms in the final modulus expression: the maximum signal is only obtained when all of the exponential terms in the summation are real.

In a more general case where all pixels are offset by a mean pixel phase value c rather than the first pixel phase value, we could state that $I = 1$ only if each element of the vector **a** has a value $\alpha_l = c + 2n\pi$ where n is an arbitrary integer. We can also express the signal explicitly as $I(\mathbf{a})$; this is a function of the vector **a**, which describes a point in an N-dimensional space. In this way, we can see that $I(\mathbf{a})$ has maximum value 1 at the origin of this space when c is zero. Furthermore, we see that there is an infinite number of points in this space at which $I(\mathbf{a}) = 1$. For example, on each of the axes, there are points where $I(\mathbf{a}) = 1$ that are equally spaced at steps of 2π. Varying the value of c is equivalent to adding a constant phase to every pixel (or equivalent to adding the piston mode to the whole pupil) and thus has no effect on the signal. We deduce therefore that there are infinite lines of $I(\mathbf{a}) = 1$ parallel to the vector $(1, 1, \ldots, 1)^T$. As the value of c has no effect on the signal, we can set this arbitrarily to zero without affecting further analysis. This is equivalent to removing the piston mode. It is also equivalent to taking the $(N-1)$ dimensional subspace including the origin in an orientation orthogonal to the direction $(1, 1, \ldots, 1)^T$. The position of the maxima in this slice would be equivalent to the positions of a scaled version of the integer lattice $\mathbb{Z}_N$, as explained in Section 3, projected along the direction $(1, 1, \ldots, 1)^T$. An illustration is provided in Figure 4.

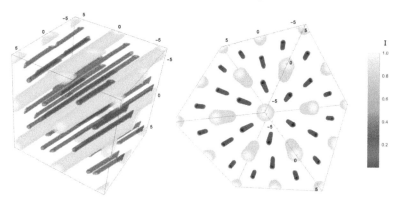

Figure 4. Illustration of the lattice geometry for intensity variation with pixel phase value. As it is not possible to represent higher order systems in a three-dimensional rendering, the example shown is for a three-pixel system. While this system does not use Walsh modes, it shows the same phenomena of piston invariance and lattice-like behaviour. The axes represent each of the pixel phase values in radians. The same volume rendering is shown from two different angles. The visible contours are set at $I = 0.01$ (blue) and $I = 0.8$ (orange) to show the positions of the zeros and the maxima, respectively. The function I is invariant with the piston mode, hence the elongation of the contours along the direction $(1, 1, 1)$. The lattice like structure of the function is apparent, in this case in the form of the hexagonal lattice. This shows that there are many different combinations of aberration mode coefficients that provide a similar well-corrected state. Analogous behaviour is found in higher dimensions for the Walsh-mode-based systems.

8. Lattice Representation of States after Removal of Piston

The removal of the piston mode is equivalent to the removal of the first row of the matrix $\mathbf{W}$ to create the reduced matrix $\mathbf{W}'$ and removing the corresponding element of $\mathbf{b}$ to obtain a reduced vector $\mathbf{b}'$ so that

$$\mathbf{a} = \mathbf{W}'^T \mathbf{b}' \tag{14}$$

As row–column products represent dot products between Walsh vectors, the following relationship is valid: $\mathbf{W}'\mathbf{W}'^T = N\mathbf{I}_{N-1}$, so Equation (14) can be inverted as

$$\mathbf{b}' = \frac{1}{N}\mathbf{W}'\mathbf{a} \tag{15}$$

We can interpret Equation (15) as a transformation from an N-dimensional vector of pixel values $\mathbf{a}$ to an $(N-1)$-dimensional vector of Walsh modal coefficients $\mathbf{b}'$.

The matrix $\mathbf{W}'$ is, however, a redundant representation, as the column space has dimension greater than its rank. This is rectified by removal of any one of the columns to create the matrix $\mathbf{W}''$; we choose arbitrarily to remove the first column. In order to maintain the form of Equation (15), we remove the first element of the vector $\mathbf{a}$ to produce a new vector $\mathbf{a}'$. From a practical perspective, this means that the pixel value a_1 is a dependent parameter determined by the other pixel values because of removal of the piston mode.

$$\mathbf{b}' = \frac{1}{N}\mathbf{W}''\mathbf{a}' \tag{16}$$

The operation performed by matrix $\mathbf{W}''$ is to map the vector $\mathbf{a}'$ to the corresponding vector $\mathbf{b}'$. Similarly, the operation of $\mathbf{W}''$ would be to transform (project) the positions of the maxima of $I(\mathbf{a}) = 1$, which were located on lines passing through a scaled integer lattice $\mathbb{Z}_N$ (as illustrated in Figure 4), to another lattice in the $(N-1)$-dimensional space spanned by $\mathbf{b}'$.

We can determine the properties of this new lattice by considering its Gram matrix, which is the matrix of the inner products between its lattice vectors [10]. The Gram matrix is hence given by

$$\mathbf{G} = \frac{1}{N}\mathbf{W}''^T\mathbf{W}'' = \frac{1}{N}\begin{pmatrix} N-1 & -1 & \cdots \\ -1 & N-1 & \cdots \\ \vdots & \vdots & \ddots \end{pmatrix} \tag{17}$$

where the factor of $1/N$ has been chosen so that the basis vectors are equivalent to Walsh functions with normalised vector magnitudes. $\mathbf{G}$ is equivalent to the Gram matrix of the so-called A^*_{N-1} lattice [10], which is an $(N-1)$-dimensional analogue of the body centred cubic (BCC) lattice in three dimensions. It follows that the maxima in the $(N-1)$-dimensional space spanned by $\mathbf{b}'$ must be located at the lattice points of a scaled A^*_{N-1} lattice.

Understanding the symmetries of this lattice thus provides an understanding of the symmetries of the function $I(\mathbf{b}')$. For example, the response of the signal around each lattice point should be identical. In other words, $I(\mathbf{b}' - \mathbf{d}_m) = I(\mathbf{b}')$ for all m, where $\mathbf{d}_m$ represents an arbitrary lattice point. As there is an infinite number of lattice points, there is an infinite combination of the $(N-1)$ Walsh coefficients that can provide the optimal correction. Furthermore, correction to a precision of $\phi_{rms} \leq \epsilon$ can be achieved by finding a setting for the adaptive correction device that places $\mathbf{b}'$ within a sphere of radius ϵ centred upon any of the lattice points.

9. Fundamental Correction Space

The lattice model allows us to define a fundamental correction space—that is, the range of $\mathbf{b}'$ we must search to find an optimal correction. This fundamental correction space is smaller than the correction space covered by all pixel values in the range 0 to 2π radians. The lattice symmetry of the function $I(\mathbf{b}')$ indicates that we need only search the Voronoi cell of the lattice in order to cover all possible states. Therefore, the search space is the Voronoi cell of the scaled A_{N-1}^* lattice, whose properties are known [10]. The position of the cell's vertices can be readily calculated. For example, the Voronoi cell of the A_3^* (or BCC) lattice is a truncated octahedron; this would be the Voronoi cell for a $N = 4$ pixel system and is illustrated in Figure 5.

Using the symmetries of the Voronoi cell, further general properties of this fundamental correction space can be derived. Moving along any of the axes from a lattice point at which $I(\mathbf{b}') = 1$, we encounter another lattice point at a distance $b = \pi$ (noting that this corresponds to the variation of a single Walsh function, the pixel values of which will be $\pm \pi$ for this value of b; see Equation (9)). Therefore, the halfway point between lattice points along the axis is at a distance $b = \pi/2$. Hence, the distance between two faces of the Voronoi cell along such an axis is π. By looking solely along the axes, one might assume that the search space is an $(N-1)$-dimensional cube of side length π, which would have a volume of π^{N-1}. However, the Voronoi cell's volume is given by

$$\pi^{N-1}\sqrt{\det \mathbf{G}} = \frac{\pi^{N-1}}{\sqrt{N}} \tag{18}$$

which is a factor $1/\sqrt{N}$ smaller than the encompassing hypercube [10]. Hence, for large numbers of pixels, the search space is considerably smaller than might be assumed if considering the pixel phases directly.

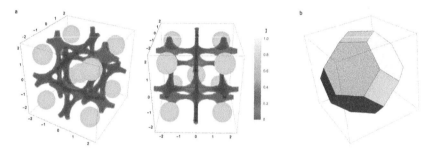

Figure 5. (a) Illustration of the lattice geometry for the fundamental correction space of a four-pixel system, which corresponds to three Walsh modes after neglecting the piston mode. The axes represent each of the Walsh coefficient values in radians. The same volume rendering is shown from two different angles. The visible contours are set at $I = 0.01$ (blue) and $I = 0.8$ (orange) to show the positions of the zeros and the maxima, respectively. The BCC lattice geometry is apparent. (b) The Voronoi cell for the A_3^* (or BCC) lattice, a truncated octahedron, is shown within an encompassing cube of side length π radians.

10. Implications of the Lattice Structure for Sensorless AO

We have shown that searching the Voronoi cell of the A_{N-1}^* lattice is sufficient to find the optimal correction in the whole Walsh coefficient space of the adaptive element. The lattice structure also means that this same cell repeats over the whole space. Consequently, if the aberration in the system can be accurately represented by a finite number of Walsh modes, then the necessary search space is finite. This contrasts with an aberration represented by a finite number of continuous modes, such as Zernike polynomials, where the search space would have to be infinite in extent to cover all possible coefficient values.

In modal sensorless AO correction schemes, a sequence of predetermined bias aberrations for fixed set of correction modes is applied to the adaptive element and the corresponding signal values are recorded. From this set of measurements, the correction aberration is estimated using an appropriately chosen optimisation algorithm. When using continuous modes, such estimation can provide accurate correction for aberrations over a limited magnitude range but usually provides poor estimation outside this range. Using Walsh modes, however, the finite search space within one Voronoi cell of the lattice structure for a fixed set of modes means that it is possible to design a correction scheme that is accurate across all possible aberrations within the Walsh mode set. For the continuous modes, the bias aberrations span a finite range, such as in the typical configuration for sensorless AO of having equal magnitude positive and negative biases for each mode. However, the same configuration for Walsh modes in effect spans an infinite range, as the bias positions are also repeated in the lattice structure across the whole coefficient space.

If a sequence of intensity measurements is taken with bias aberrations defined as each Walsh mode with an amplitude of $\pi/2$, the set of measurements is related to the Walsh transform of the pixel values (see Appendix B). The sum of these measurements is equal to the intensity after aberration correction. This has important implications for the normalisation of measurements for use in aberration estimation algorithms.

11. Neural Network for Solution of the Inverse Problem

We used the knowledge of the fundamental correction space to design estimators for a sensorless AO scheme. Specifically, the estimation process should solve the inverse problem to obtain aberration coefficients from the set of biased intensity measurements. We choose to demonstrate this with a NN estimator, whose design incorporates physical knowledge of the system. This method was chosen as it is more readily extendable to more advanced AO systems than conventional optimisation algorithms.

In this demonstration, we compare two similar NN-based methods for which the search space is defined differently. In the first case, it was assumed that each of the Walsh mode coefficients a_k can take any value $-\pi/2 < a_k \leq \pi/2$. This was equivalent to taking any point in a $(N-1)$-dimensional cube in coefficient space (we refer to this as the "hypercube cell"). In the second case, the coefficients were chosen so that they lie only within the Voronoi cell centred at the origin (we refer to this as the "primary Voronoi cell"). This primary Voronoi cell would be a sub-region of the hypercube used in the first case. Based upon the previous analysis, it was known that there would be a single point corresponding to optimum correction in the primary Voronoi cell but multiple such points in the hypercube cell.

Having multiple global optima in the search space can be detrimental when using neural networks to perform such an optimisation. This is because such ill-conditioned problems have no unique answer and thus prevent convergence of the network training. We employed a bespoke NN architecture that was developed to take advantage of the particular physical process used in the sensorless AO scheme. The overall process and the NN architecture are outlined here. More details about the NN can be found in Appendix D. In order to adequately sample the space, a biasing scheme was chosen that used $2N - 1$ measurements. This corresponded to the application of positive and negative biases of magnitude $\pi/3$ for each of the $(N-1)$ Walsh modes, excluding piston; an additional nonbiased measurement was also included. For the kth mode, we denote the negatively bias measurement as I_k^{-1}, the positively biased measurement as I_k^{+1} and the unbiased measurement as I_0.

The NN process was constructed as shown in Figure 6; a more detailed description of the architecture is given in Appendix D. The NN takes two separate sets of inputs, both of which rely on biased intensities (generated using Equation (1)): the first (*Input1*) directly uses these normalised intensity values, while the second (*Input2*) analytically processes the intensities based upon sinusoidal estimation to acquire a set of preliminary aberration coefficient estimates. The first input is passed into a convolutional neural network (CNN) followed by fully connected layers (FCL). It is then concatenated with the second input (*Input2*) and passed into fully connected layers to generate the outputs that correspond to the estimated Walsh coefficients. The rationale behind this dual input approach was that the learning task would be easier if based, in effect, on the differences between rough estimates and actual measurements rather than on the raw measurements themselves.

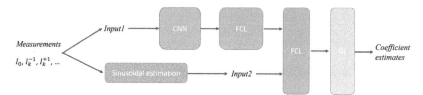

Figure 6. Outline of the NN architecture and preprocessing of data. CNN: convolutional neural network; FCL: fully connected layer; OL: output layer.

Input1 was structured in the instance of $N = 8$, as a matrix in the following form

$$Input1 = \begin{pmatrix} I_0 & I_0 & I_0 \\ I_0 & I_1^{+1} & I_1^{-1} \\ I_0 & I_2^{+1} & I_2^{-1} \\ I_0 & I_3^{+1} & I_3^{-1} \\ I_0 & I_4^{+1} & I_4^{-1} \\ I_0 & I_5^{+1} & I_5^{-1} \\ I_0 & I_6^{+1} & I_6^{-1} \\ I_0 & I_7^{+1} & I_7^{-1} \end{pmatrix} \tag{19}$$

This structure was chosen so that the CNN block could interpret known correlations from biased measurement values. Indeed, the first CNN layer was structured so that it operated first on the 3-tuples of intensity values contained in each row of *Input1*, which were each expected to depend primarily on the corresponding coefficient a_k. Further CNN layers sought to operate on correlations between the different coefficients.

The rough estimation used for *Input2* was based upon the knowledge that variations in a single Walsh mode coefficient led to a sinusoidal relationship with detected signal (see Appendix C). The value of each coefficient in *Input2* was set to the value that provided the maximum value of this sinusoidal variation. This estimation provides only a rough value for correction, as it treats each coefficient separately and hence does not deal with the coupling effect between combinations of modes.

Separate instances of this NN architecture were trained for the two scenarios. In the first case of the hypercube cell, the training and validation data were obtained by assigning a random value to each input coefficient in the range $-\pi/2 < a_k \leq \pi/2$. For the second case, the same combinations of coefficients were "wrapped" so that they lay only within the primary Voronoi cell (see Appendix D). In each case, the same calculated intensity measurements were used at the input. The difference between the two training processes lies in the coefficient labels that were used to calculate the loss function during network training—in the first case, the coefficient labels were defined throughout the hypercube cell; in the second case, the corresponding labels were wrapped into the primary Voronoi cell. Full details of the training procedure are provided in Appendix D.

Results are shown in Figure 7 for the case of $N = 8$, which corresponds to the correction of 7 Walsh modes. The loss function curves showed that only the scenario of confining the coefficient to within the primary Voronoi cell permitted the NN to converge. The mean squared phase error was reduced to 0.063 radians after correction, which corresponded to a Strehl ratio of approximately 0.6. In comparison, the loss for the hypercube case was significantly higher and the validation loss did not reduce while deviating from the training loss. This demonstrated the effectiveness of using prior knowledge about the lattice symmetries. It is expected that similar trends would be seen in scenarios with large pixel numbers, alongside more complicated NN architectures.

These results show clearly that the prior knowledge of the lattice geometry, and hence the need to confine the estimation to the primary Voronoi cell, is essential to effective training of the NN. When training using coefficients selected from the hypercube cell, the loss functions (which are related to wavefront estimation errors) do not converge sufficiently to provide good aberration correction. This is attributed to the existence of multiple optimal solutions within the hypercube cell that complicate the training process. However, this problem is avoided when using a training set where the labels were "wrapped" and the resulting labels were within the primary Voronoi cell, and thus only one optimal combination of Walsh mode coefficients is present.

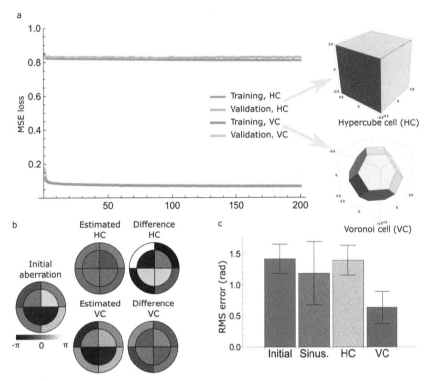

Figure 7. (a) NN training and validation loss functions as mean square error (MSE) for the scenarios where the coefficient labels were defined throughout the hypercube cell (HC) or the primary Voronoi cell (VC) for $N = 8$. The insets on the right show a schematic representation in three dimensions the difference between the two types of cells. (b) An illustrative example of correction of an initial aberration consisting of 8 pixels, equivalently 7 polar Walsh modes. The residual error of the VC-based correction was far lower than that of the HC based correction. (c) Statistical summary of correction results from the NN validation set: Initial—distribution of initial input aberrations; Sinus—after estimation using sinusoidal model; HC—after correction using the hypercube cell method; VC—after correction with the Voronoi cell method. Error bars show the standard deviation of the distribution.

12. Conclusions

The insight provided by the mathematical link between Walsh mode AO and lattices informs the design of efficient sensorless AO schemes. This is relevant to the solution of the inverse problem of how to estimate the aberration coefficients from metric measurements with different applied bias aberrations. The new insights are particularly important when using NNs to solve this inverse problem, as otherwise we suffer from the challenges caused by having multiple solutions for a given set of metric measurements, which requires more complicated networks.

Although the Walsh modes are pixelated, when sufficient pixels are used, they can provide a suitably accurate representation of low order aberrations. While continuous modes, such as the Zernike polynomials, are commonly used, they are not guaranteed to provide a good fit to high-order aberrations in the system, nor to the correction device, which may well be pixelated in nature. The results presented here have relevance not just for correction of low-order aberrations but also for high-order scattering compensation, where previously schemes have been based around control of Walsh modes (or similar) [15–17].

The analysis presented in this paper was based around a simple AO focussing system. However, as the overall lattice geometry arises from the nature of the aberration representation, and not the AO system or the optimisation metric, a similar repetitive lattice structure and primary Voronoi cell will hold for any other AO system using Walsh modes as the basis. These results should therefore have relevance to any application of AO using pixelated correction devices.

Author Contributions: Conceptualization: M.J.B.; methodology: M.J.B. and Q.H.; software: M.J.B., Q.H. and Y.X.; validation: Q.H., J.C., R.T. and Y.X.; investigation: M.J.B., Q.H. and Y.X.; writing—original draft preparation: M.J.B. and Q.H.; writing—review and editing: R.T., J.C. and Y.X.; visualization: M.J.B. and Q.H.; supervision: M.J.B.; funding acquisition: M.J.B. All authors have read and agreed to the published version of the manuscript.

Funding: This research was funded by the European Research Council, grant number 695140 (AdOMiS).

Conflicts of Interest: The authors declare no conflict of interest.

Abbreviations

The following abbreviations are used in this manuscript:

AO	Adaptive optics
DM	Deformable mirror
SLM	Spatial light modulator
NN	Neural networks
SNR	Signal to noise ratio
AE	Adaptive element
CNN	Convolutional neural network
FCL	Fully connected layers
WT	Walsh transform

Appendix A. Observations Based upon the Lattice Geometry

Various connections can be made between the operation of the Walsh mode-based AO system and the lattice representation outlined above. The following points address certain relationships between variations of individual pixels and the Walsh functions.

- A single pixel variation is put into effect by a combination of all N Walsh modes, which can be obtained using Equation (5). As is clear from this equation, the coefficients in the vector **b** are themselves given by the elements of a Walsh mode.
- As we have removed the first (piston) mode, we can see that each reduced Walsh mode (i.e., without its first element) is a vector of length $N - 1$ that points in the direction corresponding to the variation of a single pixel. As these vectors each include all the $N - 1$ basis vectors in equal magnitude (± 1), then these directions must correspond to some of the body diagonals between opposite vertices through the centre of the fundamental cubic cell that encloses the Voronoi cell.
- For N pixels, there are $2N$ body diagonals that correspond to single pixel variations (counting positive and negative directions separately). For $N > 4$, this is less than the total number of body diagonals, which is 2^N. Hence, for $N > 4$, there are diagonals that do not correspond to single pixel variations, but would involve multiple changing pixels.
- The "kissing number" τ is a characteristic of a lattice that indicates how many nearest neighbours there are to any lattice point. The kissing number for the A_n^* lattice is $\tau = 2n + 2$ (for $n \geq 2$), so for A_{N-1}^* we expect $\tau = 2N$ (p155 of [10]). This is equal to the number of body diagonals noted above as corresponding to single pixel variations.

These single pixel variations correspond to the kissing directions, which are the closest spacing between lattice points.

- Equation (8) shows how I varies with single pixel modulation. However, removal of the piston mode leads to some differences. Ensuring zero piston means that all pixels are modulated, but $N-1$ pixels are shifted by the same value $-\xi$, whereas the single desired pixel will be shifted by the value $(N-1)\xi$. The peak-to-peak amplitude would be $\psi = N\xi$.
- The mean square amplitude of such a mode must be given by

$$\frac{\left[(N-1)\xi^2 + (N-1)^2\xi^2\right]}{N} = (N-1)\xi^2 = \frac{N-1}{N^2}\psi^2 \quad (A1)$$

Hence, the rms phase related to the peak-to-peak phase by $\phi_{rms} = \frac{\sqrt{N-1}}{N}\psi$.

- The distance between these closest lattice points is equivalent to the rms phase required to shift one pixel so that it is 2π radians different to the others. Setting $\psi = 2\pi$ gives $\phi_{rms} = \frac{2\pi\sqrt{N-1}}{N}$. This varies inversely with $\sqrt{N}$, such that as N increases the spacing between closest lattice points reduces. This is to be expected, as increasing N means a smaller pixel size and hence a smaller rms phase for a given phase variation of a single pixel.
- Along these kissing directions, the minimum signal is obtained when the single pixel is π out of phase with the other $N-1$ pixels. This leads to a signal minimum of

$$I = \frac{(N-2)^2}{N^2} \quad (A2)$$

which tends to 1 as N increases and, correspondingly, as the size of a single pixel decreases.

Hence, we can also use this lattice description to show that the signal varies only weakly in these kissing directions (corresponding to single pixels), whereas adjustment of Walsh modes provides more robust measurement. This confirms the results presented in Figure 3 using single pixel and modal variations. These properties are closely related to the extent of the Voronoi cell in these kissing directions. The Voronoi cell is in effect narrower in the kissing directions than in others. In general, larger signal modulations are obtained if one samples the function in the directions where the Voronoi cell has a larger extent.

Appendix B. Sensorless AO and the Walsh Transform

Suppose we take a sequence of biased measurements, where we apply biases of each Walsh mode with a single positive amplitude of $\pi/2$ radians. This is equivalent to shifting the measurement point from the body centre of the $(N-1)$-dimensional Voronoi cell to the centre of each $(N-2)$-dimensional facet centred along the positive axis of the corresponding Walsh mode. Due to the lattice structure, the facet in the positive direction is homologous to the facet in the negative direction. Therefore, by choosing this bias amplitude, we are simultaneously sampling both the positive and negative bias positions. When biasing the kth mode, we are increasing the (unknown) coefficient β_k by $\pi/2$. The complex pixel value of that mode is therefore

$$\exp\left(i\left\{\beta_k + \frac{\pi}{2}\right\}W_k^N[l]\right) = \exp\left(i\frac{\pi}{2}W_k^N[l]\right)\exp\left(i\beta_k W_k^N[l]\right)$$
$$= iW_k^N[l]\exp\left(i\beta_k W_k^N[l]\right) \quad (A3)$$

where we have exploited the fact that the Walsh function has values only of ± 1. We can now see that the kth biased signal measurement is given from Equation (7) by

$$I_k = \left|\frac{i}{N}\sum_{l=1}^{N}\left\{\exp\left(i\sum_{k=0}^{N-1}\beta_k W_k^N[l]\right)W_k^N[l]\right\}\right|^2 = \left|\frac{i}{N}\sum_{l=1}^{N}x_l W_k^N[l]\right|^2 \quad (A4)$$

where we have defined x_l to represent the complex pixel values so that:

$$x_l = \exp\left(i \sum_{k=0}^{N-1} \beta_k W_k^N[l]\right) \quad \text{(A5)}$$

The Walsh transform (WT) X_k of a sequence x_l of length N is defined conventionally as [8]:

$$X_k = \frac{1}{N} \sum_{l=0}^{N-1} x_l W_k^N[l] \quad \text{(A6)}$$

We can therefore derive

$$I_k = |X_k|^2 \quad \text{(A7)}$$

In other words, we see that the set of biased intensity measurements is the modulus squared of the kth component of the WT of the complex pixel values x_l. As the first Walsh function is piston (hence all pixels have value 1), it is related to the first element of the WT X_k. As biasing with piston has no effect on the measurement, this is the same as the unbiased measurement in our system. Hence, the unbiased measurement (corresponding to the centre of the Voronoi cell) along with the $N - 1$ biased measurements correspond to a full set of intensity measurements. This set of intensity measurements is known as the spectral density of the WT of the set of complex pixel values. In practice, we cannot measure the complex values of X_k directly, but we are able to quantify the Walsh spectral density through the intensity measurements with the applied bias modes.

We now derive a relationship using the fundamental properties of the WT. In [8], we find an equivalent of Parseval's theorem for WTs:

$$\sum_{k=0}^{N-1} |X_k|^2 = \frac{1}{N} \sum_{l=1}^{N} |x_l|^2 \quad \text{(A8)}$$

From the definition of x_l, it is clear that its modulus is equal to one. Hence, we find that

$$\sum_{k=0}^{N-1} |X_k|^2 = 1 \quad \text{(A9)}$$

Equivalently, we find that the sum of the $N - 1$ biased measurements and the one unbiased measurement must add to one.

The importance of this result is that in a real experiment the measured signal is not actually normalised to one, but there is an unknown "brightness" that is a function of numerous experimental variables. This would be a multiplying factor in the expressions for I, which we have chosen to omit from the analysis for simplicity. We can use this result to obtain the "brightness" in a way that is independent of the input aberration, as it is simply the sum of the $N - 1$ biased measurements and the unbiased measurement.

Appendix C. Estimation of Coefficients Using Simple Sinusoidal Model

In this appendix, we present a method for rough initial estimation of individual Walsh mode coefficients. Let us redefine Equation (7) by including a factor A that is equivalent to an unknown maximum intensity

$$I = A \left| \frac{1}{N} \sum_{l=1}^{N} \exp\left(i \sum_{k=0}^{N-1} \beta_k W_k^N[l]\right) \right|^2 \quad \text{(A10)}$$

The signal varies with a single modal coefficient β_j as

$$I(\beta_j) = A \left| \frac{1}{N} \sum_{l=1}^{N} \exp\left(i \sum_{k=0; k \neq j}^{N-1} \beta_k W_k^N[l]\right) \exp\left(i\beta_j W_j^N[l]\right) \right|^2$$

$$= A \left| \frac{1}{N} \sum_{l=1}^{N} C[l] \exp\left(i\beta_j W_j^N[l]\right) \right|^2 \tag{A11}$$

where the effects of the modes contained within the first exponential term have been subsumed into the complex coefficients $C[l]$.

We introduce a useful relationship, which takes advantage of the binary (± 1) values of the Walsh functions [9]:

$$\exp\left(i\beta_j W_j^N[l]\right) = \cos\left(\beta_j W_j^N[l]\right) + i\sin\left(\beta_j W_j^N[l]\right) = \cos\beta_j + iW_j^N[l]\sin\beta_j \tag{A12}$$

This leads to

$$I(\beta_j) = A \left| \frac{1}{N} \sum_{l=1}^{N} C[l] \left\{\cos\beta_j + iW_j^N[l]\sin\beta_j\right\} \right|^2 \tag{A13}$$

The exact form of this solution will depend upon $C[l]$ and hence on the other Walsh coefficients. However, the only terms in β_j that can arise from the modulus square term are of the form of $\cos^2\beta_j$, $\cos\beta_j\sin\beta_j$, or $\sin^2\beta_j$, which can all be expressed as sinusoidal terms of argument $2\beta_j$. Hence, we deduce that $I(\beta_j)$ must be given by the form

$$I(\beta_j) = A\left[p + q\cos\left(2\beta_j + \zeta\right)\right] \tag{A14}$$

where p, q and ζ depend on all values $\beta_{k \neq j}$. Let us simplify Equation (A14) to the form $I(\theta_j) = U + V\cos(2\theta_j)$, where we have defined $2\theta_j = 2\beta_j + \zeta$.

Consider applying a bias bW_j^N so that the biased measurement $I_+ = U + V\cos(2\theta_j + 2b)$. Then, we apply the negative bias $-bW_j^N$ so that the biased measurement $I_- = U + V\cos(2\theta_j - 2b)$. We also use the unbiased measurement $I_0 = U + V\cos(2\theta_j)$. If we use a bias of $b = \frac{\pi}{3}$, then we can obtain through elementary operations

$$\theta_j = \frac{1}{2}\tan^{-1}\left[\sqrt{3}\frac{I_+ - I_-}{I_+ + I_- - 2I_0}\right] = \beta_j + \frac{\zeta}{2} \tag{A15}$$

The value of θ_j obtained through Equation (A15) was used as the initial rough estimate provided to the NN as *Input2*. The error between this estimate and the actual coefficient is given by $\beta_j - \theta_j = -\zeta/2$.

Appendix D. Description of the Neural Network Training and Network Architecture

The training data consisted of 2^{20} (about one million) simulated samples, and 2^7 (128) samples were used for validation to avoid overfitting. Both sets of data were generated for the case of $N = 8$, corresponding to the correction of seven Walsh modes (excluding piston). The total number of training datasets was chosen to provide a sufficient representation of the seven-dimensional search space.

For each training dataset, seven values were randomly generated corresponding to the coefficients of seven polar Walsh modes (β_k). The coefficients followed a uniform distribution over the range of $-\pi/2$ to $\pi/2$. These coefficients were used as labels during network training in the case of the hypercube cell.

The sum of the product between each coefficient with its corresponding unit Walsh mode formed the phase aberration $\Psi(\mathbf{r})$ according to Equation (3). The phasor of the complex field was calculated by averaging the complex field over the pupil. By subtracting the angle of the mean phasor of $\Psi(\mathbf{r})$ across the pupil (Equation (A16)) and wrapping the phase back within the range of $\pm\pi$, the aberration $\Psi'(\mathbf{r})$ was calculated to be equivalent to $\Psi(\mathbf{r})$ and within the primary Voronoi cell.

$$\Psi'(\mathbf{r}) = \arg\left(\exp\left\{i\left[\Psi(\mathbf{r}) - \arg\left(\frac{1}{\pi}\int \exp[i\Psi(\mathbf{r})]\, d\mathbf{r}\right)\right]\right\}\right) \quad (A16)$$

Using the orthogonality of Walsh modes, the new coefficients β_k' of the equivalent aberration $\Psi'(\mathbf{r})$ (Equation (A17)) could be calculated. β_k' were used as the labels when training the network with a confined searching space.

$$\beta_k' = \frac{1}{\pi}\int \Psi'(\mathbf{r})\omega_k(\mathbf{r})d\mathbf{r} \quad (A17)$$

For each set of data, two phase biases per Walsh mode were introduced. The bias amplitudes were chosen to be $-\pi/3$ and $\pi/3$. For each introduced bias phase, the intensity signal (I_- and I_+) was computed using the Equation (1). The same equation was used to calculate the signal when no bias phase was introduced (I_0). From previous discussions, the phase modulation of mode 1 (piston) would have no effect on the signal and thus excluded from the collection of signal readings. The total of 15 signal readings (14 biased and 1 unbiased readings) were used as the input (*Input1*) of the network.

In addition, from our previous discussion, the signal varied with mode coefficients in a period of π. A good approximation to the coefficients of each mode (β_k'') was obtained using Equation (A17). These approximations were also used as the separate input (*Input2*) to the two networks we trained. Figure A1 shows a few sets of β_k, β_k' and β_k'', which derived from the same initial aberration.

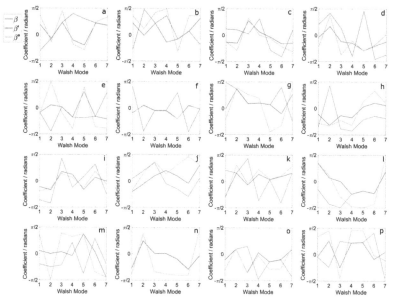

Figure A1. (**a**–**p**) 16 sets of randomly selected β, β' and β'' derived from the same aberration, shown in blue, red and yellow plots, respectively. In some cases, β'' (estimation using sinusoidal model) closely resembled β' (such as in case (**f**)) while in some other cases, β'' could be different to β' (such as in (**e**) and (**n**)). There were also a small proportion of cases where β was identical to β' (such as (**n**)), which corresponded to the initial aberration being within the primary Voronoi cell.

The full diagram of the NN architecture is shown in Figure A2. The network was built using TensorFlow Keras. For the first CNN layer, the padding was chosen to be "valid" and strides equalled to (1,1). For each of the second to fourth CNN layers, the padding was chosen to be "same" and strides equalled to (1,1). Following each of the second to the fourth convolutional layers was a maxpooling layer with a pool size of 2 × 1. For all the layers (except the output layer), the nonlinear activation was chosen to be "tanh". The activation of the last output layer was linear. The initializer of all the kernels was glorot uniform. The loss function was mean squared error (MSE). The optimizer was Adam.

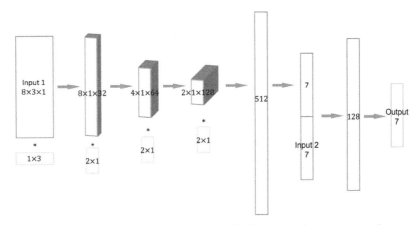

Figure A2. Full neural network architecture. The blue boxes were the two inputs to the network. The yellow boxes were the trainable kernels of the CNN with the corresponding dimensions as shown. The orange boxes were the internal layers of the CNN and dense layers in the later stages. The green box represents the output of the CNN.

References

1. Booth, M.J. Adaptive optical microscopy: The ongoing quest for a perfect image. *Light Sci. Appl.* **2014**, *3*, e165. [CrossRef]
2. Ji, N. Adaptive optical fluorescence microscopy. *Nat. Methods* **2017**, *14*, 374–380. [CrossRef] [PubMed]
3. Hampson, K.M.; Turcotte, R.; Miller, D.T.; Kurokawa, K.; Males, J.R.; Ji, N.; Booth, M.J. Adaptive optics for high-resolution imaging. *Nat. Rev. Methods Prim.* **2021**, *1*, 68. [CrossRef] [PubMed]
4. Booth, M.J.; Neil, M.A.A.; Juškaitis, R.; Wilson, T. Adaptive aberration correction in a confocal microscope. *Proc. Natl. Acad. Sci. USA* **2002**, *99*, 5788–5792. [CrossRef] [PubMed]
5. Hu, Q.; Wang, J.; Antonello, J.; Hailstone, M.; Wincott, M.; Turcotte, R.; Gala, D.; Booth, M.J. A universal framework for microscope sensorless adaptive optics: Generalized aberration representations. *APL Photonics* **2020**, *5*, 100801. [CrossRef]
6. Facomprez, A.; Beaurepaire, E.; Débarre, D. Accuracy of correction in modal sensorless adaptive optics. *Opt. Express* **2012**, *20*, 2598–2612. [CrossRef] [PubMed]
7. Saha, D.; Schmidt, U.; Zhang, Q.; Barbotin, A.; Hu, Q.; Ji, N.; Booth, M.J.; Weigert, M.; Myers, E.W. Practical sensorless aberration estimation for 3D microscopy with deep learning. *Opt. Express* **2020**, *28*, 29044–29053. [CrossRef] [PubMed]
8. Beauchamp, K. *Walsh Functions and Their Applications*; Nutrition, Basic and Applied Science; Academic Press: Cambridge, MA, USA, 1975.
9. Wang, F. Wavefront sensing through measurements of binary aberration modes. *Appl. Opt.* **2009**, *48*, 2865–2870. /AO.48.002865. [CrossRef] [PubMed]
10. Conway, J.H.; Sloane, N.J.A. *Sphere Packings, Lattices and Groups*, 3rd ed.; Grundlehren der mathematischen Wissenschaften, A Series of Comprehensive Studies in Mathematics; Springer: New York, NY, USA, 1999; Volume 290.
11. Booth, M.J. Wave front sensor-less adaptive optics: a model-based approach using sphere packings. *Opt. Express* **2006**, *14*, 1339–1352. [CrossRef] [PubMed]
12. Antonello, J.; Verhaegen, M.; Fraanje, R.; van Werkhoven, T.; Gerritsen, H.C.; Keller, C.U. Semidefinite programming for model-based sensorless adaptive optics. *J. Opt. Soc. Am. A* **2012**, *29*, 2428–2438. [CrossRef] [PubMed]
13. Weisstein, E.W. Hadamard Matrix—From Wolfram MathWorld. Available online: https://mathworld.wolfram.com/HadamardMatrix.html (accessed on 18 July 2022).
14. Sloane, N.J.A. Hadamard Matrices. Available online: http://neilsloane.com/hadamard/ (accessed on 18 July 2022).
15. Tang, J.; Germain, R.N.; Cui, M. Superpenetration optical microscopy by iterative multiphoton adaptive compensation technique. *Proc. Natl. Acad. Sci. USA* **2012**, *109*, 8434–8439. [CrossRef] [PubMed]

16. Park, J.H.; Sun, W.; Cui, M. High-resolution in vivo imaging of mouse brain through the intact skull. *Proc. Natl. Acad. Sci. USA* **2015**, *112*, 9236–9241. [CrossRef] [PubMed]
17. Kong, L.; Cui, M. In vivo neuroimaging through the highly scattering tissue via iterative multi-photon adaptive compensation technique. *Opt. Express* **2015**, *23*, 6145–6150. [CrossRef] [PubMed]

Article

Real-Time Correction of a Laser Beam Wavefront Distorted by an Artificial Turbulent Heated Airflow

Alexey Rukosuev *, Alexander Nikitin, Vladimir Toporovsky, Julia Sheldakova and Alexis Kudryashov

Institute of Geosphere Dynamics (IDG RAS), 119334 Moscow, Russia; nikitin@activeoptics.ru (A.N.); topor@activeoptics.ru (V.T.); sheldakova@nightn.ru (J.S.); kud@activeoptics.ru (A.K.)
* Correspondence: alru@nightn.ru

Abstract: This paper presents a FPGA-based closed-loop adaptive optical system with a bimorph deformable mirror for correction of the phase perturbation caused by artificial turbulence. The system's operating frequency of about 2000 Hz is, in many cases, sufficient to provide the real-time mode. The results of the correction of the wavefront of laser radiation distorted by the airflow formed in the laboratory conditions with the help of a fan heater are presented. For detailed consideration, the expansion of the wavefront by Zernike polynomials is used with further statistical analysis based on the discrete Fourier transform. The result of the work is an estimation of the correction efficiency of the wavefront distorted by the turbulent phase fluctuations. The ability of the bimorph adaptive mirror to correct for certain aberrations is also determined. As a result, it was concluded that the adaptive bimorph mirrors, together with a fast adaptive optical system based on FPGA, can be used to compensate wavefront distortions caused by atmospheric turbulence in the real-time mode.

Keywords: adaptive optics; bimorph mirror; fast adaptive optical system; turbulence; wavefront sensor; Zernike polynomials; spectral analysis

1. Introduction

Laser radiation, propagating in the Earth's atmosphere and beyond, allows us to solve the following tasks:

- Crypto-protected information transmission [1];
- Organization of the optical communication channels in free space [2];
- Recharging batteries of drones [3] and low-orbit satellites [4];
- Destruction of space debris [5];
- Therefore, on.

It is known [6] that air layers with different temperatures lead to the formation of turbulent refractive index changes along the propagation of the radiation. Passing through such layers, the laser beam wavefront (WF) acquires additional phase incursions, which leads to the degradation of the beam as it propagates from the source to the receiver. This limits the scope of application of the laser systems operating in a real atmosphere. One of the ways to solve this problem is to use an adaptive optical system (AOS) that is capable of compensating for the phase nonuniformity of the wavefront in real time.

At the same time, the system should have sufficient performance. As shown, for example, in [6], the frequency of phase fluctuations caused by atmospheric turbulence rarely exceeds 100 Hz. To compensate for such aberrations, a discrete AOS with a correction frequency of at least 1000 Hz (frames per second) is required. It is quite difficult to provide such a high stable performance using a conventional PC, since in addition to measuring the wavefront and calculating the voltages vector, the system should transmit information to the control unit of the deformable mirror. It is much more efficient to use a FPGA for these purposes, in which a full cycle of wavefront correction will be implemented. The use of

FPGAs will allow you to achieve a frequency that will be sufficient to correct the wavefront in real time.

It should be noted that in all previously mentioned tasks, the use of an AOS is potential, and in each case, a separate study of its applicability is required. In particular, it is necessary to solve different extra problems such as obtaining a reference signal, combating scintillations, etc.

Currently, there is a huge interest in this topic in the world. For example, in the article [7], a 9 km horizontal maritime links experiment was performed through adaptive correction. In the article [8], the issues of laser radiation correction without the use of a wavefront sensor on a 2.3 km long path with subsequent conjugation with fiber are considered. The article [9] describes an attempt to use adaptive correction in relation to the global-scale optical clock network to improve the residual instability along the 113 km path.

Before starting experiments on an open-air route, in our case, it was advisable to investigate the capabilities of the system in laboratory conditions. For this purpose, a laboratory setup was assembled, where the heated airflow from the fan heater acted as a source of distortion of the wavefront. Fourier analysis [10] applied to the data coming from the Shack–Hartmann wavefront sensor (WFS) [11,12] showed [13] that the spectral power density of the turbulent airflow is close to Kolmogorov's within the bandwidth of about 60 Hz. This parameter indicates the proximity of the laboratory conditions to the real atmosphere.

The next step in the research was the use of the expansion of the wavefront by Zernike polynomials [14]. This work is a continuation of the studies of the wavefront aberrations caused by the heat flow of the air described in [15]. Now, a similar analysis was carried out, but in the case of correction of the wavefront of laser radiation using an AOS system operating at the frequency of 2000 Hz. As a result, the efficiency of the system was evaluated, as well as the ability of the used bimorph mirror to correct for the specific aberrations.

2. The Fast Adaptive Optical System

The AOS includes a wavefront corrector (deformable mirror), a wavefront sensor and a control system. To achieve the specified correction speed in the experiments, a system controlled by the FPGA was used.

2.1. Deformable Mirror

A key element of any AOS is a wavefront corrector (WFC). It determines the ability of the system to compensate for specific aberrations, and also affects the performance of the system as a whole. When choosing the WFC, the previously measured Fried parameter [16] was taken into account, which in experiments turned out to be equal to 10 mm [17]. The measurements were carried out according to the method described in [18], where the Fried parameter was determined from the mutual dispersion of the oscillations of two focal spots of the lens array of the Shack–Hartmann wavefront sensor. The focal spots were chosen to be centrally symmetric with respect to the center of the beam. To obtain a more reliable result, several pairs of points were selected and the result of calculating the Fried parameter was averaged over these calculations.

Based on the values of the Fried parameter, it was proposed to use a bimorph mirror [19,20] as a WFC. This mirror has sufficient speed. The frequency of the first resonance is 8.3 kHz; the amplitude-phase frequency response (Bode diagram) is shown in Figure 1. The phase-frequency response of the mirror becomes equal to 90 degrees at a frequency of about 8.2 kHz, while the peak of the amplitude resonance is at 8.3 kHz.

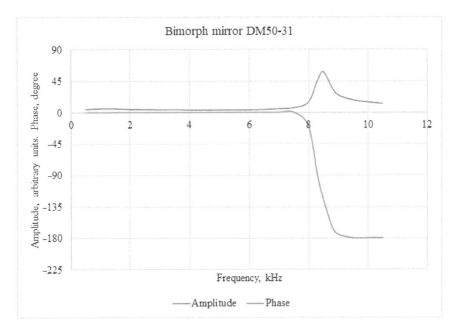

Figure 1. Bode diagram for bimorph deformable mirror.

The structure of the electrodes consists of three 8 mm wide rings, which should be enough to correct for the WF with the Fried parameter of 10 mm. The photo of the mirror and the structure of the electrodes are shown in Figure 2. Table 1 shows the main characteristics of the mirror.

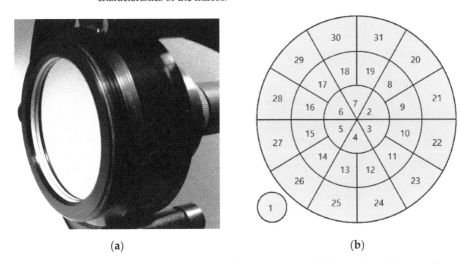

(a)   (b)

Figure 2. Bimorph mirror used in experiments: (**a**) Mirror photo; (**b**) electrodes structure. The numbers indicate the channels of the mirror. The first electrode, conventionally shown in the lower-left corner, has the dimensions of the mirror aperture and is used to control the defocusing of the wavefront.

Table 1. Main parameters of the bimorph mirror.

Parameter	Value
Clear aperture	50 mm
Electrodes number	31
Control voltage range	−200 V ± 300 V
Maximal stroke	±10 µm
First resonant frequency	8.3 kHz
Coating	Protected silver
Size	Ø 70 mm × 68 mm
Weight	320 g

2.2. Shack–Hartmann Wavefront Sensor

To ensure the system operating frequency of the 2000 Hz, a wavefront sensor based on a fast camera was used. The performance improvement was achieved by using a part of the image. Thus, in the experiments we used a resolution of 480 × 480 pixels, which allowed us to increase the frame rate to 4000 Hz. The main parameters of the wavefront sensor are presented in Table 2

Table 2. Main parameters of the wavefront sensor.

Parameter	Value
Sensor	Alexima LUX19HS
Spectral bandwidth	350–1100 nm
Dynamic Range (Tilt)	±50λ
Accuracy of measurements	λ/90
Frame rate	2500 fps @ 1920 × 1080
	~4000 fps @ 480 × 480
Interface	Fiber Optic 40 Gb/s
Lenslet array focal length	12 mm
Number of working sub-apertures	20 × 20
Input light beam size	4.8 × 4.8 mm
Resolution	8 bit

2.3. Adaptive Optical System Control Loop

To ensure the fast operation of the AOS, the control loop was made using FPGA. The system requires (1) preloading of a reference—a set of coordinates of WFS focal spots [21]—to which the coordinates of the real focal spots will be pulled up using WFC; and (2) preloading of WFC response functions. Based on the analysis of the image coming from the WFS, the FPGA calculates a vector of corrective voltages, which is then applied to the mirror electrodes. To achieve a high speed of operation in the system, a phase-conjugated algorithm was used.

Since the bimorph deformable mirror cannot reproduce tilts, virtual tilts response functions are introduced into the system to exclude them. In fact, the voltages are also calculated for tilts, but they are not applied anywhere.

In the experiments, the AOS operated with a correction frequency of 2000 Hz (frames per second) in the closed-loop mode. To obtain such a high-speed performance, FPGA Arria V GZ processed the image bytes coming from the WFS camera 'on the fly' (as they arrive from the camera), which made it possible to calculate the vector of control voltages by the end of the frame reception. Since the sensor camera operated at a frequency of 4000 Hz, it took 250 microseconds. One hundred and fifty microseconds were required to transmit the voltage vector to the adaptive mirror Control Unit, fifty microseconds to load the DAC, and the pause required for the mirror to change its shape lasted fifty microseconds. The internal structure of the FPGA is shown in Figure 3, and the corresponding time diagram of its operation is shown in Figure 4.

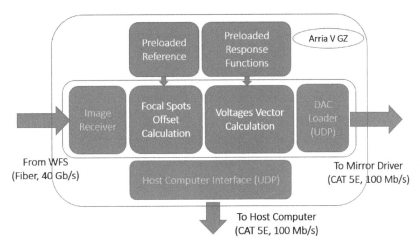

Figure 3. FPGA inner structure.

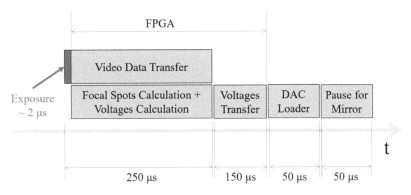

Figure 4. FPGA timing diagram. The total time of one closed cycle is 500 microseconds, which corresponds to an operating frequency of 2000 Hz.

3. Experimental Setup

To correct for the WF, a laboratory installation was used, as shown in Figure 5.

The source of the radiation is a laser diode coupled to an optical fiber. The wavelength is chosen as 650 nm to facilitate the alignment. The collimating lens forms a parallel beam with a diameter of 50 mm. A fan heater is installed on the path of propagation of the laser beam, the heated airflow of which crosses the laser beam, creating the turbulence. A laser beam with a distorted wavefront hits a WFC-bimorph mirror and is reflected from it in the direction of the WFS. A flat mirror installed between the WFC and the WFS is used to reduce the dimensions of the installation. Part of the beam in front of the WFS branches off to the far-field indicator formed by a long-focus lens and a CMOS camera with a small pixel size (3.75 microns) for a more detailed visualization of the image. The operation of the system is controlled by the FPGA, which performs all the functions necessary for the correction of the wavefront: it receives an image from the WFS camera, calculates the vector of correcting voltages and transmits this vector to the WFC amplifier unit (Mirror CU). The PC in this configuration performs only the functions of controlling the operation of the FPGA and the functions of monitoring the correction process.

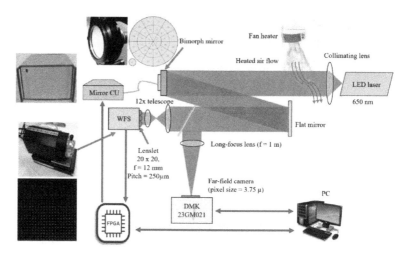

Figure 5. Adaptive optical system experimental setup. WFS—wavefront sensor; mirror CU—mirror control unit.

4. Processing the Results of the Experiment

Before starting the work, studies of distortions of the wavefront of the laser beam caused by the influence of a heated air stream were carried out. Figure 6 shows the spectral power density of the process obtained on the basis of statistical processing of a series of recorded fluctuations in the coordinates of the focal spot of the WFS lens array. The straight line $f^{(-5/3)}$ corresponds to the Kolmogorov spectrum. The sampling duration of the focal spot coordinates was 10 s (at a frequency of 2000 Hz, 20,000 samples were recorded), which provided a resolution along the frequency axis of 0.1 Hz.

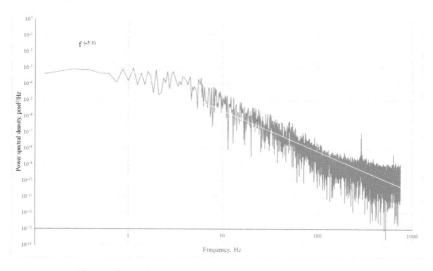

Figure 6. Spectral power density of the oscillation of the X coordinate of the WFS lens array focal spot.

The quality control of the correction was carried out on the far field image [22]. With the specified parameters of the laboratory setup (beam diameter of 50 mm, lens focus length of the far field zone indicator of 1 m and radiation wavelength of 650 nm), the diffraction-limited diameter of the spot in the lens focus was about 32 microns. Correction of the laser radiation wavefront makes it possible to obtain a focal spot diameter in the far field at the level of 9 pixels, which, with a pixel size of 3.75 microns, corresponds to 34 microns. Thus, the diameter of the spot in the far field zone is close to the diffraction limit and, consequently, the correction quality is quite good. Figure 7 shows images of the focal spot in the far field. The upper pictures were obtained in the absence of correction, while the lower picture shows the intensity distribution in the presence of wavefront correction. It should be noted that in the absence of correction, the intensity of the focal spot was too low, so the upper row of images was obtained with a longer exposure (85 µs vs. 57 µs for the corrected case). The image of the far field is close to the diffraction one and practically does not change its shape during the correction procedure.

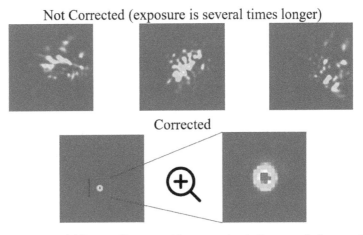

Figure 7. Far-field images. Top row—without correction; bottom row—during correction.

Another way to assess the quality of the correction is the residual error. Figures 8 and 9 show the change in residual RMS over time for the case of correction and the absence of correction. Figure 8 represents a complete set of aberrations and graph on Figure 9 was obtained by excluding the tilts.

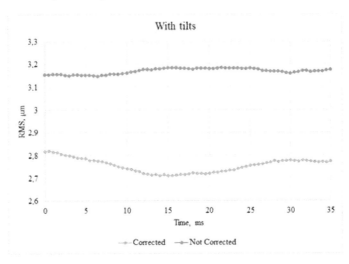

Figure 8. Residual RMS changes over time; full set of aberrations.

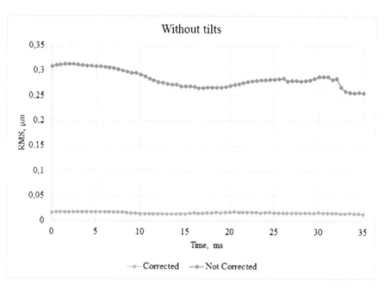

Figure 9. Residual RMS changes over time, without tilts.

For a more detailed consideration of the quality of correction and to obtain quantitative results, a transition was made to the spectral analysis of each of the modes of the wavefront decomposition by Zernike polynomials. The data-processing algorithm was as follows.

- A sample of the offset coordinates of the focal spots of the lens array of the WFS with a duration of 10 s was recorded. This made it possible to achieve a resolution along the frequency axis of 0.1 Hz. At a frequency of 2000 Hz, a total of 20,000 values of coordinate offsets of each focal spot were recorded
- The transition was carried out from sampling by coordinates to sampling by the coefficients of the wavefront expansion by Zernike polynomials. In this work, we used a set of 24 Zernike polynomials in Wyant indices [23] (1 and 2—tilts, 3—defocus, 4 and 5—astigmatism, 6 and 7—coma, 8—spherical, etc.).
- Using the Fourier transform, the transition from the time domain to the frequency domain was carried out for sampling each Zernike polynomial;
- The power spectral density was calculated for each mode;
- Further, by integrating the spectral power density, the spectral energy was calculated.

If we integrate the spectral power density for each Zernike polynomial, we can get a graph of the spectral energy (Figure 10). The graphs at a certain frequency value go into saturation, which indicates an insignificant contribution of higher-frequency components to the total signal energy. The frequency of transition of graphs to saturation can be taken as the bandwidth occupied by one Zernike polynomial or another.

Figure 11 shows a diagram of the dependence of the frequency band occupied by one aberration or another. The graph corresponds to one of the consequences of Taylor's hypothesis [24], according to which the rate of change of high-order aberrations is faster.

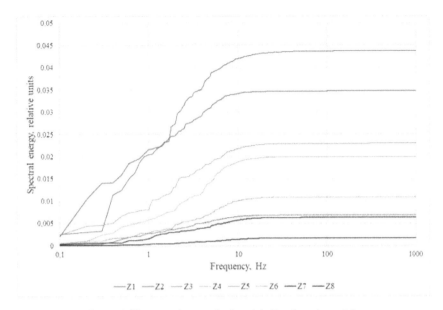

Figure 10. The spectral energy for first eight Zernike polynomials.

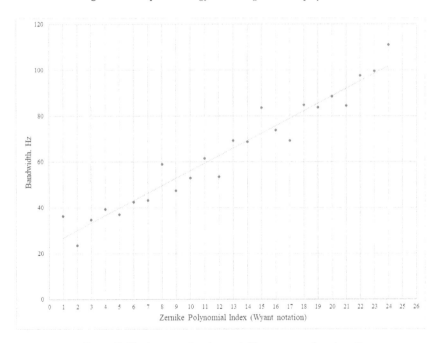

Figure 11. The frequency band occupied by one or another aberration.

The spectral energy saturation amplitude from Figure 10 for each Zernike polynomial is shown on Figure 12.

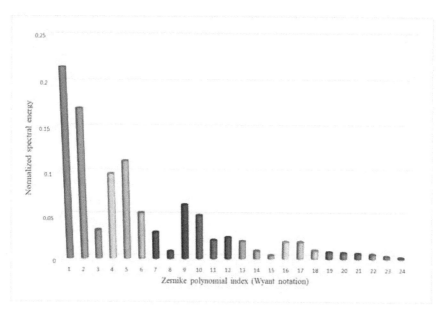

Figure 12. The spectral energy of wavefront aberrations before correction.

The residual aberrations are quite small, so Figure 13 represents a diagram of the uncompensated aberrations, expressed as a percentage of the entry level.

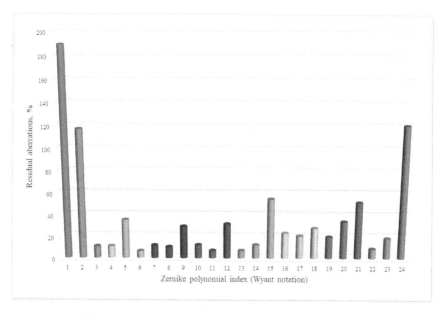

Figure 13. Residual aberrations after correction.

For greater clarity, Figure 14 shows the same diagram, but expressed in decibels.

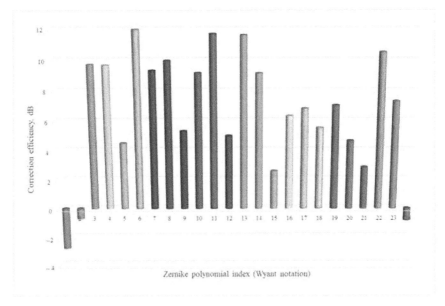

Figure 14. Correction efficiency for each Zernike mode.

5. Discussion

Based on the diagram on Figure 14, we can make the following suggestions.

1. The combination of FPGA performance and a bimorph wavefront corrector in an adaptive optical system allows correction of artificially created turbulence in real time. The correction frequency in the experiments was chosen to be equal to 2000 Hz. Such a stable correction frequency is almost impossible to obtain using a conventional PC. The PC, unlike the FPGA, performs I/O at the driver level, thereby increasing the time of the closed correction cycle. FPGA exchanges data between external devices (wavefront sensor and corrector control unit) directly. In addition, the FPGA performs parallel processing of information, which has significant limitations in the case of using a PC.
2. The speed of the AOS controlled by the FPGA made it possible to analyze the effectiveness of aberration correction up to the 23rd Zernike polynomial, whose bandwidth is about 100 Hz, in detail.
3. The bimorph WFC does not have the ability to correct for the slopes of the WF. Accordingly, the correction efficiency of the first two Zernike polynomials is negative, i.e., there is an increase in the amplitude of the initial slopes. To eliminate the slopes, it is necessary to use either a separate beam position stabilization system (see, for example, [25]), or to install a mirror in a tip–tilt mount. In this case, the virtual slopes used in the experiment become real and the voltages calculated during operation are applied to the control drives of the tip–tilt mount.
4. WFC used in the experiments has three rings of electrodes and cannot reproduce a spherical aberration of the third order (polynomial # 24), since this aberration has four extremums (max–min). To reduce the aberration # 24, a higher spatial resolution WFC is required.
5. The aberrations from 3 to 23 are compensated well enough by this WFC. However, because the initial amplitude of the polynomial # 24 is small compared to other aberrations, the undercompensating of this aberration can be neglected.

6. Conclusions

We demonstrated a closed-loop adaptive optical system with the bimorph deformable mirror, Shack–Hartmann wavefront sensor and FPGA controller that can efficiently correct for the wavefront aberrations caused by the artificial turbulence. The total speed of the system operation was equal to 2000 Hz. It should be noted that the bimorph deformable mirror corrects for the wavefront aberrations caused by the flow of the heated air quite well—the residual error of the phase fluctuations was reduced by more than 10 times and an almost diffraction-limited focal spot was obtained. Certainly, when using such a type of corrector, it is necessary to additionally apply a system to stabilize the position of the beam in space.

Author Contributions: Conceptualization, A.K., A.R.; methodology, A.K., A.R.; software, A.R.; validation, A.K., A.N., J.S.; formal analysis, A.N., A.R., V.T.; investigation, A.R., A.N.; resources, A.K.; data curation, A.R., V.T.; writing—original draft preparation, A.R.; writing—review and editing, A.N., J.S., V.T.; visualization, A.N., A.R.; supervision, A.K.; project administration, A.K. All authors have read and agreed to the published version of the manuscript.

Funding: The research was carried out within the state assignment of Ministry of Science and Higher Education of the Russian Federation (theme # 122032900183-1).

Data Availability Statement: The data presented in this study are available on request from the corresponding author.

Conflicts of Interest: The authors declare no conflict of interest. The funders had no role in the design of the study; in the collection, analyses, or interpretation of data; in the writing of the manuscript, or in the decision to publish the results.

References

1. Huang, Q.; Liu, D.; Chen, Y.; Wang, Y.; Tan, J.; Chen, W.; Liu, J.; Zhu, N. Secure free-space optical communication system based on data fragmentation multipath transmission technology. *Opt. Express* **2018**, *26*, 13536–13542. [CrossRef] [PubMed]
2. Vorontsov, M.; Weyrauch, T.; Carhart, G.; Beresnev, L. Adaptive optics for free space laser communications. In *Lasers, Sources and Related Photonic Devices*; OSA Technical Digest Series (CD), LSMA1; Optical Publishing Group: Washington, DC, USA, 2010. [CrossRef]
3. Lu, M.; Bagheri, M.; James, A.P.; Phung, T. Wireless charging techniques for UAVs: A review, reconceptualization, and extension. *IEEE Access* **2018**, *6*, 29865–29884. [CrossRef]
4. Landis, G.A.; Westerlund, H. *Laser Beamed Power—Satellite Demonstration Applications*; NASA Contractor Report 190793, IAF-92-0600; NASA: Washington, DC, USA, 1992.
5. Bennet, F.; Conan, R.; D'Orgeville, C.; Dawson, M.; Paulin, N.; Price, I.; Rigaut, F.; Ritchie, I.; Smith, C.; Uhlendorf, K. Adaptive optics for laser space debris removal. *Proc. SPIE* **2012**, *8447*, 844744. [CrossRef]
6. Tatarskii, V.I. *Wave Propagation in a Turbulent Medium*; McGraw-Hill: New York, NY, USA, 1961; 304p.
7. Rui, W.; Yukun, W.; Chengbin, J.; Xianghui, Y.; Shaoxin, W.; Chengliang, Y.; Zhaoliang, C.; Quanquan, M.; Shijie, G.; Li, X. Demonstration of horizontal free-space laser communication with the effect of the bandwidth of adaptive optics system. *Opt. Commun.* **2019**, *431*, 167–173. [CrossRef]
8. Weyrauch, T.; Vorontsov, M. Free-space laser communications with adaptive optics: Atmospheric compensation experiments. Free-Space Laser Communications. *Opt. Fiber Commun. Rep.* **2004**, *2*, 247–271. [CrossRef]
9. Shen, Q.; Guan, J.-Y.; Ren, J.-G.; Zeng, T.; Hou, L.; Li, M.; Cao, Y.; Han, J.-J.; Lian, M.-Z.; Chen, Y.-W.; et al. 113 km Free-Space Time-Frequency Dissemination at the 19th Decimal Instability. *arXiv* **2022**, arXiv:2203.11272. [CrossRef]
10. Brigham, E.O. *The Fast Fourier Transform and Its Applications*; Prentice Hall: Upper Saddle River, NJ, USA, 1988; 448p.
11. Hardy, J. Active optics: A new technology for the control of light. *Proc. IEEE* **1978**, *66*, 651–697. [CrossRef]
12. Cao, G.; Yu, X. Accuracy analysis of a Hartmann–Shack wave front sensor operated with a faint object. *Opt. Eng.* **1994**, *33*, 2331–2335. [CrossRef]
13. Rukosuev, A.L.; Nikitin, A.N.; Sheldakova, Y.V.; Kudryashov, A.V.; Belousov, V.N.; Bogachev, V.A.; Volkov, M.V.; Garanin, S.G.; Starikov, F.A. Smart adaptive optical system for correcting the laser wavefront distorted by atmospheric turbulence. *Quantum Electron.* **2020**, *50*, 707–709. [CrossRef]
14. Born, M.; Wolf, E. *Principles of Optics: Electromagnetic Theory of Propagation, Interference and Diffraction of Light*, 1st ed.; Pergamon Press: London, UK; New York, NY, USA; Paris, France, 1959; 986p.
15. Rukosuev, A.; Nikitin, A.; Belousov, V.; Sheldakova, J.; Toporovsky, V.; Kudryashov, A. Expansion of the Laser Beam Wavefront in Terms of Zernike Polynomials in the Problem of Turbulence Testing. *Appl. Sci.* **2021**, *11*, 12112. [CrossRef]

16. Fried, D.L. Optical Resolution Through a Randomly Inhomogeneous Medium for Very Long and Very Short Exposures. *J. Opt. Soc. Am.* **1966**, *56*, 1372–1379. [CrossRef]
17. Belousov, V.; Bogachev, V.; Volkov, M.; Garanin, S.; Kudryashov, A.; Nikitin, A.; Rukosuev, A.; Starikov, F.; Sheldakova, Y.; Shnyagin, R. Investigation of spatial and temporal characteristics of turbulent-distorted laser radiation during its dynamic phase correction in an adaptive optical system. *Quantum Electron.* **2021**, *51*, 992–999. [CrossRef]
18. Sarazin, M.; Roddier, F. The ESO Differential Image Motion Monitor. *Astron. Astrophys.* **1990**, *227*, 294–300.
19. Kudryashov, A.; Samarkin, V.; Aleksandrov, A. Adaptive optical elements for laser beam control. *Proc. SPIE* **2001**, *4457*, 170–178. [CrossRef]
20. Toporovskii, V.; Skvortsov, A.; Kudryashov, A.; Samarkin, V.; Sheldakova, Y.; Pshonkin, D. Flexible bimorphic mirror with high density of control electrodes for correcting wavefront aberrations. *J. Opt. Technol.* **2019**, *86*, 32–38. [CrossRef]
21. Nikitin, A.; Sheldakova, J.; Kudryashov, A.; Borsoni, G.; Denisov, D.; Karasik, V.; Sakharov, A. A device based on the Shack-Hartmann wave front sensor for testing wide aperture optics. *Proc. SPIE* **2016**, *9754*, 97540K. [CrossRef]
22. Kudryashov, A.; Rukosuev, A.; Nikitin, A.; Galaktionov, I.; Sheldakova, J. Real-time 1.5 kHz adaptive optical system to correct for atmospheric turbulence. *Opt. Express* **2020**, *28*, 37546–37552. [CrossRef]
23. Goodwin, E.; Wyant, J. *Field Guide to Interferometric Optical Testing*; SPIE Press Book: Bellingham, WA, USA, 2006; 114p, ISBN 0-8194-6510-0.
24. Taylor, G.I. The Spectrum of Turbulence. *Proc. Roy. Soc.* **1938**, *164*, 476–490. [CrossRef]
25. Saathof, R.; Breeje, R.; Klop, W.; Doelman, N.; Moens, T.; Gruber, M.; Russchenberg, T.; Pettazzi, F.; Human, J.; Mata Calvo, R.; et al. Pre-correction adaptive optics performance for a 10 km laser link. In *Free-Space Laser Communications XXXI*; SPIE: Bellingham, WA, USA, 2019; Volume 10910, pp. 325–331. [CrossRef]

Article

On-Demand Phase Control of a 7-Fiber Amplifiers Array with Neural Network and Quasi-Reinforcement Learning

Maksym Shpakovych [1], Geoffrey Maulion [1], Alexandre Boju [1,2], Paul Armand [1], Alain Barthélémy [1], Agnès Desfarges-Berthelemot [1] and Vincent Kermene [1,*]

[1] XLIM, Faculté des Sciences et Techniques, Université de Limoges-CNRS UMR n°7252 123 Ave. A. Thomas, F-87060 Limoges, France; maksym.shpakovych@xlim.fr (M.S.); geoffrey.maulion@xlim.fr (G.M.); alexandre.boju@xlim.fr (A.B.); paul.armand@xlim.fr (P.A.); alain.barthelemy@xlim.fr (A.B.); agnes.desfarges-berthelemot@xlim.fr (A.D.-B.)

[2] CILAS Ariane Group, 8 Avenue Buffon, CS16319, CEDEX 2, F-45063 Orléans, France

* Correspondence: vincent.kermene@xlim.fr

Abstract: We report a coherent beam combining technique using a specific quasi-reinforcement learning scheme. A neural network learned by this method enables the tailoring and locking of a tiled beam array on any phase map. We present the experimental implementation of on-demand phase control by a neural network in a seven-fiber laser array. This servo loop needs only six phase corrections to converge to the desired phase set at any profile, with a bandwidth higher than 1 kHz. Moreover, we demonstrate the dynamical feature of adaptive phase control, performing sequences of controlled phase sets. It is the first time, to the best of our knowledge, that an actual array of seven-fiber amplifiers has been successfully phase-locked and controlled by machine learning.

Keywords: coherent beam combining; neural network; adaptive optics; laser beam array; deep learning

Citation: Shpakovych, M.; Maulion, G.; Boju, A.; Armand, P.; Barthélémy, A.; Desfarges-Berthelemot, A.; Kermene, V. On-Demand Phase Control of a 7-Fiber Amplifiers Array with Neural Network and Quasi-Reinforcement Learning. *Photonics* **2022**, *9*, 243. https://doi.org/10.3390/photonics9040243

Received: 14 March 2022
Accepted: 1 April 2022
Published: 6 April 2022

Publisher's Note: MDPI stays neutral with regard to jurisdictional claims in published maps and institutional affiliations.

Copyright: © 2022 by the authors. Licensee MDPI, Basel, Switzerland. This article is an open access article distributed under the terms and conditions of the Creative Commons Attribution (CC BY) license (https://creativecommons.org/licenses/by/4.0/).

1. Introduction

Coherent beam combining (CBC) of multiple emitters represents a key versatile technique in providing high average power or high-energy short pulses while maintaining beam quality [1]. The CBC architectures are designed to handle the laser power distributed over a set of amplification channels arranged in parallel. Due to thermal effects and mechanical instabilities, each channel phase of the piston type must be adjusted over time to maintain the combining efficiency and wavefront quality of the combined beam. There are two methods of performing the combining step, such as the tiled-aperture and filled-aperture techniques. In the first configuration, the amplified beams are placed side by side to form a kind of large synthetic pupil and are then coherently overlapped in the far field. In the second configuration, they are superimposed by splitters or by a diffractive optical element (DOE) in the near field to obtain a single high-power beam. The tiled-aperture arrangement offers the opportunity to dynamically shape the synthetic wavefront by tuning the piston phase of each element of the array to a desired value. This dynamic shaping could be useful particularly for compensation of phase aberration due to atmospheric perturbations in the context of directed energy production [2,3]. CBC was also recently investigated to shape the far field pattern of a high-power beam array. In particular, T. Hou et al. numerically validated the generation of orbital angular momentum (OAM) laser beams in a tiled-aperture architecture [4]. In 2021, M. Veinhard et al. demonstrated OAM beam shaping by tailoring the phase of 61 beams in the femtosecond regime [5]. These specific modes, which preserve their ring intensity profile during propagation, are of interest in many areas such as particle manipulation and free-space propagation. Moreover, real-time control of intensity shape at focus by CBC at a high-power level can optimize the performance of material processing.

An active coherent combining device with fiber amplifiers is based on a master oscillator power amplifier (MOPA) configuration with multiple parallel fiber amplifiers that undergo internal and environmental perturbations. The phase fluctuation compensation at the output of the fiber array is realized by electro-optic modulators which command comes from direct measurements of the current output phase state [6,7], or by correcting the phase in an iterative way to optimize a given parameter [8–11]. In the latter case, the loop performing the phase correction includes an optimization algorithm such as the popular stochastic parallel gradient-descent (SPGD) method or the alternating projection (AP) method [12–14]. With the SPGD method, as the beams count increases, the correction bandwidth drops significantly. The AP method, on the contrary, is well suited to the phase-lock of a wide beam array at the expense of a large number of detectors. A third method, based on neural network and deep learning, was recently investigated.

Among the many applications of neural networks (NN) in optics, few of them recently published dealt with CBC [15–22]. In most cases, the papers reported numerical studies. Some contributions investigated NN for direct, one-step, phase recovery of the beam array from scattered patterns through a diffuser [17] or through a diffractive optical element DOE [20]. In the latter case, an NN with only two layers provided accurate phase recovery but in a limited phase error range. Despite it being trained in a limited range, once applied in a feedback system for phase correction, the technique was able to compensate for a full range $[-\pi, \pi]$ of random initial phase errors and to reach phase-locking. It required approximately 40 iterations on average to lock a 9×9 array, which was demonstrated to be ten times faster than SPGD optimization. A reinforcement learning method was also considered as a second option for beam combining with NN [15,19,21]. In a first experiment with a two-fiber interferometer [15], the authors demonstrated the technique could be as efficient as a standard PID (proportional integrator differentiator) controller or as SPGD. Previous simulations on deep reinforcement learning with a deep deterministic policy gradient have used the far field pattern as input to the NN. Locking of the phase was shown to require 6 to 12 iterations for a 7-beam array [19]. However, they raised issues regarding scalability for large arrays in particular due to the dimensionality of the training data set, a loss in accuracy and the duration of the training. The approach offers the additional capability of tailoring the array far field, such as, for the generation of orbital angular momentum beams (OAM) [18]. In a recent publication [22], we proposed a third option, called quasi-reinforcement learning (QRL). Training of the NN for phase-locking was carried out specifically for operation in a loop with a given number of iterations. Simulations and a proof of principle experiment demonstrated efficient and fast (six iterations) phase-locking of a 100-beam array.

In this paper, we report first a new version of this machine-learning scheme that provides access to instantaneous tailoring and locking of a tiled beam array on any phase map. Then, we present experiments of its implementation in a seven-fiber laser array. It is the first time, to the best of our best knowledge, that an actual array of fiber amplifiers has been successfully phase-locked and controlled by an NN.

In the following paragraph, we first briefly remind the reader of the principle of the approach, as detailed in [22]. Then, we describe the improved version of the NN implemented in the QRL process, which allows real-time adaptive changes of the desired phase map for the laser beam array. Finally, we present an experimental phase control from the QRL approach in the dynamic environment of a fiber laser array. This shows that the iterative phase-locking process converges to any static or dynamic desired phase relationship with a correction loop bandwidth over 1 kHz.

2. Neural Network in a Phase Reduction Loop with Quasi-Reinforcement Learning

The system we have previously proposed to control the phase of a laser beam array [22] (laser fields of complex amplitude z with unknown phases) is described on Figure 1. It is composed of (i) a diffuser for mapping individual phases into intensity through scattering, (ii) a photo-detector array, which converts optical intensity into voltage, (iii) an NN, which processes the electrical signal and provides correction commands to an array of phase modulators. The NN serves to perform the inverse of the transformation achieved by the diffuser. From sparse samples (measurements b^2) of the scattered intensity pattern, it predicts a value $\tilde{z}$ of the individual laser fields in the array. Knowledge of the presumed phase set $\arg(\tilde{z})$ and of the desired phase set $\arg(z_d)$ then permits computation of the correction $= \arg(z_d) - \arg(\tilde{z})$, which serves as a command to the phase modulators. The high performance of the scheme, as demonstrated numerically and in a proof of principle experiment, relies on its specific QRL training. It consists in an optimization of the NN parameters, considering the looped operation of the system for a fixed given number of iterations T. For each round in the loop, an optimization is achieved in order to obtain the highest reward, i.e., the lowest difference between the phases after correction and the desired phases. QRL also bears a role in the learning of a recurrent neural network, although with some peculiarities. First experiments [22] showed that, unlike NN learned for direct (one-step) phase retrieval [18,20], the NN, specifically trained for phase correction in an error reduction loop, remains efficient and accurate for an array with a large number of beams (100), and for correction of phase angle on the full circle $[-\pi, +\pi]$. To preserve accuracy, the total number of iterations in the loop during training must be empirically determined, as it evolves slightly owing to the array size and to the number of intensity samples in the diffraction pattern. Most of the time it was close to $T = 6$. Once in operation, the trained NN adjusts the initial distorted phase front onto the desired one after a number of corrections less than, or equal to maximum of six.

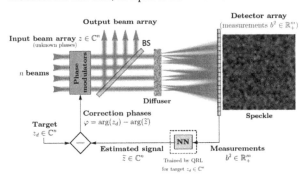

Figure 1. Principle of the system for phase-locking a coherent beam array with a neural network. In a preliminary step, quasi-reinforcement learning (QRL) trains the NN specifically for working in a feedback loop and for setting the array output to a given target phase chart. BS denotes beam splitter.

3. Target Adaptive NN with QRL Process

With the previous NN version [22], the laser beam array could be locked onto the in-phase state or any other arbitrary target phase set. However, the NN must be trained with the desired target phase set which makes a fast change of target unlikely due to the duration of the training. This explains the reason behind our proposal of implementing a target adaptive neural network (TANN) in the QRL scheme to circumvent this drawback. With this new version, the target phase set can be changed on-demand during laser system operation.

The idea is to build the network TANN that will compute the set of parameters of the NN for use in the phase-lock loop. TANN takes the vector of target phases as an input and returns the weights of the NN. Each time one modifies the desired phase profile, the NN parameters are computed again. The calculation is extremely fast (matrix vector product) and thus offers almost real-time adaptive wavefront shaping. The new adaptive phase-locking and phase-profiling system can be schematically described as shown in Figure 2.

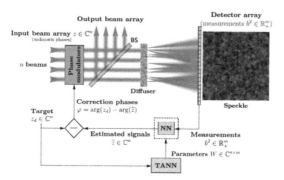

Figure 2. Feedback loop with target adaptive neural network TANN that computes the weights of the NN embedded in the loop, at each change of the target.

TANN takes as an input, a vector $z_d \in \mathbb{C}^n$ of laser fields with target phases and returns the set of parameters that is used to define the correction model for the given target. We recall that in [22] we used $NN(b) = W_2(W_1 b + \beta_1) + \beta_2$ as a correction model fed by the square root of the measurements b^2, where the set of parameters were $W_1 \in \mathbb{R}^{4n \times m}$, $W_2 \in \mathbb{R}^{2n \times 4n}$, $\beta_1 \in \mathbb{R}^{4n}$, $\beta_2 \in \mathbb{R}^{2n}$ for n beams and $m > n$ measurements. In this context, TANN should return a real vector of dimension $4nm + 8n^2 + 6n$, which is then split into several parts to define $W_{1,2}$, $\beta_{1,2}$.

This means that TANN itself has a minimum of $O(n^3)$ parameters to train. This fact requires a reduction in the number of parameters in a correction NN as much as possible. Note, that the NN in [22] is a simple affine transform $Wb + \beta$, where $W = W_2 W_1$ and $\beta = W_2 \beta_1 + \beta_2$. This smaller form decreases the number of parameters in the NN model to $2nm + 2n$. It was also observed empirically that bias β did not have a great impact on the NN's correction capability. Let us consider a new correction model of the form $NN(b) = Wb$. However, instead of using the real matrix $W \in \mathbb{R}^{2n \times m}$, which computes real and imaginary parts separately, we change it to a fully complex form $W \in \mathbb{C}^{n \times m}$. The reason behind why this smaller model was not used in [22], but had similar numerical properties, was that it required more time to train the parameters, which represents an important factor when working with 100 beams. The architecture of TANN is a simple linear map (U) from the vector of desired laser fields set $z_d \in \mathbb{C}^n$ to the vector of NN parameters, the output of which is reshaped into a matrix $TANN(z_d) = \text{Reshape}(Uz_d)$, where Reshape : $\mathbb{C}^{mn} \to \mathbb{C}^{n \times m}$ and trainable parameters $U \in \mathbb{C}^{nm \times n}$. The learning process is similar to [22] and is presented in Algorithm 1, where the reward function is a resemblance parameter between the actual array phase $arg(z)$ and the computed recovered array phase $arg(\tilde{z})$. It is defined as:

$$R(z, \tilde{z}) = \frac{|\langle z, \tilde{z} \rangle|^2}{\langle |z|, |\tilde{z}|^2 \rangle} \qquad (1)$$

In which the maximum equals 1, if and only if $arg(z) = arg(\tilde{z})$ reaches up to a constant. In the framework of laser phase-locking, $R(z, \tilde{z})$ is equal to the phasing quality Q, also called combining efficiency, which measures how close the controlled array wavefront is to uniformity. It is usually assumed that in practice an RMS deviation of $\lambda/30$ is a very good value, which corresponds to $Q = 0.96$ [23]. Therefore, this value fixes the minimum reward to reach during the training of the TANN.

As with the same concept seen in [22], the NN, which now depends on the target, computes a correction as a complex vector instead of a vector of phases. To accelerate the learning, we use a batch of targets $z_d \in \mathbb{C}^{N \times n}$ and signals $z \in \mathbb{C}^{N \times P \times n}$, where N and P denote positive natural numbers. The batch of the form $z \in \mathbb{C}^{N \times P \times n}$ means that we generate P initial signals to correct for each of the N targets during training. Note, that N and P are set to 1 in the Algorithm 1 to simplify the notation.

Algorithm 1: Quasi-reinforcement learning algorithm for TANN

Input: Measurement model: $\mathbb{C}^n \to \mathbb{R}_+^m$, reward function $R : \mathbb{C}^n \times \mathbb{C}^n \to [0,1]$
Output: Trained target adaptive neural network TANN: $\mathbb{C}^{mn \times n} \times \mathbb{C}^n \to \mathbb{C}^{n \times m}$

1. Initialize network TANN with random initial weights $U \in \mathbb{C}^{mn \times n}$
2. Set reward $r = 0$;
3. While reward $r < 0.96$ do
4. Generate a vector $z \in \mathbb{C}^n$ of random signals
5. Generate a vector $z_d \in \mathbb{C}^n$ of target signals
6. Repeat T times
   a. Measure intensities square root $b \in \mathbb{R}_+^m$ of z by M.
   b. Compute matrix $W \in \mathbb{C}^{n \times m}$ for z_d by TANN to define NN.
   c. Compute recovered field $\tilde{z} \in \mathbb{C}^n$ from amplitudes b by NN.
   d. Compute reward $r = R(z, \tilde{z})$.
   e. Update parameters of TANN to maximize r.
   f. Perform a phase correction $z = z \cdot e^{i(\arg(z_d) - \arg(\tilde{z}))}$.
7. Return trained TANN

4. Simulations

Simulations were performed for $N = 1024$, $P = 256$, $T = 8$ with a maximum of 5000 learning epochs. Signals z and targets z_d were generated as complex vectors with uniformly distributed phases on $[-\pi, \pi]$ and unit amplitudes. The initial values in $U \in \mathbb{C}^{mn \times n}$ were distributed by standard normal law. The step 6a of Algorithm 1 was performed by means of a mathematical model, instead of a direct usage of the experimental setup, which accelerated the learning process significantly. The mathematical model could be either a transmission matrix model TM or another neural network, which was referred to as NN-G in [22]. Computations were conducted on a computer using Windows 10 OS with GPU—NVIDIA GTX 1660 Ti, CPU—AMD Ryzen 5 3600 X 6-Core Processor and RAM—32 GB. To implement and train the TANN model, TensorFlow 2.5.0 library was used together with Python 3.7 language. TensorFlow encapsulates the interaction with GPU, thus we made no additional effort for parallelization. No multicore parallelization was required. Moreover, MATLAB graphical program was created to interact with the experimental setup. This program used one process for this goal.

As a particular example similar to the experiments reported below, we show in Figure 3a the evolution of the reward during training in the case of a 7-beam array with 70 measurements in the scattered pattern. The reward evolves quickly and continuously toward its maximum value in about 100 epochs. This means that the phasing quality reaches its maximum at any desired phase profile. The training required about 13 s. The phase correction process using this trained TANN shows (Figure 3b) that an average of only three iterations was enough to reach the 0.96 reward limit in a noiseless numerical study.

To obtain a full picture regarding the capabilities of TANN, several additional information slices are presented in Figure 4. It was numerically observed that in order to achieve a sufficiently high reward, say $r > 0.96$, there is a minimal required ratio m/n for different n. When the beam count varies from 4–20, the required m/n ratio increases from 4–12. Thus, it is important to show a minimal required ratio m/n for different n to achieve a sufficiently high reward. Different TANNs were trained for the various number of beams $n \in \{4, 6, 8, 10, 12, 14, 16, 20\}$ and the different ratios between the number of measurements and the number of beams $m/n \in \{2, 4, 6, 8, 10, 12, 14, 16, 18, 20\}$. The maximal achievable reward was recorded and visualized as a heat map in Figure 4a, with the corresponding relative training time shown in Figure 4b. The maximal achievable reward is obtained by solving 1000 phase correction problems with different targets for each combination of n and

m/n, and computing 95% quantile of the rewards at the last correction. This statistic reveals the minimal reward, which was obtained during the solving of 95% of test problems.

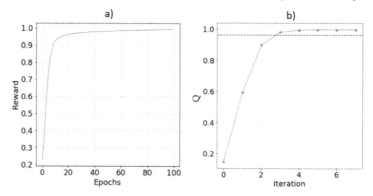

Figure 3. (a) Reward evolution during TANN training for 7 beams and 70 detectors with 8 corrections steps T. (b) Average evolution of the quality factor Q according to the phase correction iterations of the trained system (100 random initial phase sets, 7 beams, 70 detectors). The red dotted line shows the 96% threshold. On average, only 3 steps of correction are required to reach the threshold phasing quality.

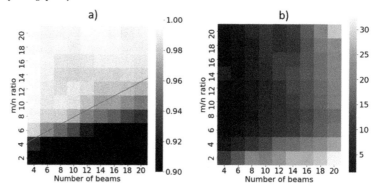

Figure 4. (a) Heat maps of the maximal achievable mean reward in grey scale and (b) its required relative training time. The red line in (a) approximates the separation line for which $r = 0.96$. The relative time on (b) is computed by dividing a learning time in seconds for each n and m/n by the minimal time to obtain GPU invariant information. The minimal time required by the GPU used in this paper was 13 s.

The red line in Figure 4a reveals the dependency between n and m/n to obtain $r = 0.96$ and is defined as $f(n) = \frac{n}{2} + 1$. This gave us information about the minimal number of measurements needed to obtain $r \geq 0.96$, which was $m = \frac{n^2}{2} + n$.

5. Experiments

We applied TANN associated with quasi-reinforcement learning to the phase-locking of a seven-amplifier laser system. As a conventional CBC configuration, the setup (Figure 5) comprised a master oscillator (MO/ CW semiconductor laser @1064nm) seeding seven parallel polarization maintaining (PM) fiber amplifiers. Their inputs were equipped with fiber-coupled LiNbO3 electro-optic phase modulators (EOM) and their outputs, once collimated by microlenses (µlens), formed a compact 1D array of laser beams (250 µm beam waist and 500 µm pitch) in a tiled-aperture arrangement (Figure 5). We used a master diode laser delivering 1064 nm radiation because most of the components used to split and modulate the light feeding the amplifier array were already in our stock and designed to operate at this popular wavelength. The wavelength choice does not impact the working principle of the investigated technique. The master laser delivered about 80 mw of polarized light. Each individual output of the double-stage polarization maintaining the fiber amplifier array was limited to about 1 W of collimated polarized laser light by the available pump power. A beam splitter (BS) split the laser array output into a power fraction and a control fraction for the phase-locking loop. The adaptive phase correction loop contains a phase sensing module made of a ground glass diffuser [14,17] which achieved interferences between the individual beams on a 1D-photodetector array. Only sparse samples of the interference pattern were collected and served as a phase to intensity encoding. We used here only 70 intensity measurements from non-adjacent and periodically spaced pixels of the photodetector array. These data fed the digitizing and processing unit. It comprised the AD/DA converters, and the QRL-learned TANN that first computed the NN to be used in the loop. The TANN received the target phase chart, which could be changed on-demand, from a computer or any other external device. The processing unit delivered the phase corrections to apply to the seven electro-optic modulators. The far field of the BS main output was displayed on a camera with a positive lens for observation and performance analysis (not shown in Figure 5).

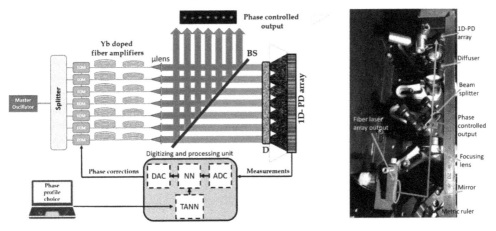

Figure 5. Left, setup of the 7-fiber amplifier array used in the reported experiments on-demand phase control using NN. The master oscillator was a semiconductor laser. EOM denotes LiNbO3 electro-optic modulator and the double-stage Ytterbium-doped fiber amplifiers were polarization maintained with 1W output power, μlens stands for microlens array, BS for beam splitter, D for diffuser. Right, photograph of the 1D-array output and of the phase analysis module.

The learning step of the TANN requires a large amount of training data. Because the experimental generation of suitable data requires a long period of time, we attained the training data by computation, using the measured transmission matrix (TM) of the scattering device that maps phase into intensity [14,24,25]. Based on the TM knowledge, we further generated a large number of training data for the TANN quasi-reinforcement learning. We set $T = 8$ as the number of correction loops in the QRL process. That number results from a previous numerical study and appears to offer a good trade-off between speed and accuracy. Optimization of the TANN parameters typically required a minimum of 100 Epochs of 256 couples of phase/intensity and 1024 target phase batches to reach a reward R of 99%. Figure 6 shows a typical evolution of the reward R versus the number of epochs during the TANN learning process with the data from the experimental TM.

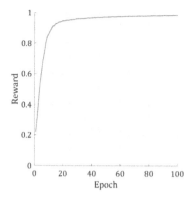

Figure 6. Reward evolution during the TANN learning process for the 7-fiber laser array with 8 correction steps T.

Once TANN was trained, we used it to compute the NN embedded in the feedback loop for phase-locking the laser array. The NN quickly and efficiently locked the laser system to the in-phase state as shown in Figure 7, despite the standing phase fluctuations in the various amplifier arms. The laser exhibited the expected far field pattern (Figure 7a), very similar in shape and magnitude to the theoretical one for an in-phase beam array (Figure 7b).

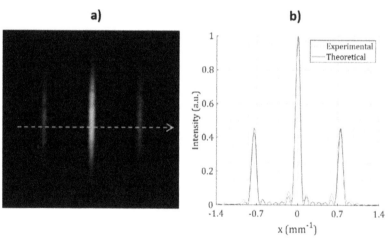

Figure 7. (a) Experimental far field of the 7-fiber laser array locked in-phase. (b) Experimental and theoretical profiles of the phase-locked fiber laser array.

The NN phase correction process locked the laser system with a measured coherent combining efficiency of ~93%, derived from the signal of a photodiode located in the center of far field. This corresponds to less than $\lambda/20$ RMS residual deviation from a perfectly uniform discrete wavefront in the beam array.

A photodiode measured the on-axis peak intensity in the array far field. To quantify the phase-locking stability of the laser system, we recorded 10 million samples of its signal during 2.8 s, (Figure 8 in-phase locking case). The samples were further analyzed to plot their probability density for the OFF (open) and ON (closed) state, respectively. When the feedback loop was open, the signal probability density (black curve in Figure 8b) covered a medium and widely spread voltage range. On the contrary, when the servo is ON (red trace), the histogram shows a sharp peak at a higher voltage (0.93) which corresponds to the average combining efficiency, associated with a 1.2% standard deviation. This demonstrates that the NN-based phase control system offers an efficient and stable locking of the fiber laser array output. The power spectral density (PSD) related to the same photodiode signal is given in Figure 8c. It shows that the servo loop corrected the phase fluctuations of the combined beam array up to 1.5 kHz, while the servo loop operated at 11 kHz frequency, limited by the speed of the loop controller (Ni PXIe-1071). The analysis of numerous on/off servo transitions shows that the average number of phase corrections to reach an efficient phase-locking level is about 6, which is quite low although slightly larger than the number derived from noiseless numerical simulations.

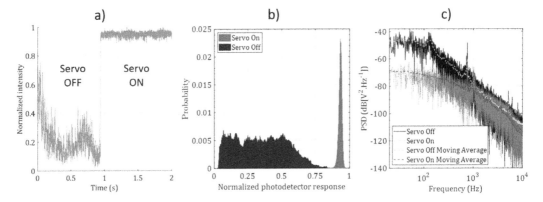

Figure 8. (a) Normalized evolution of the combined beam power detected by a photodiode located on the far field center when the NN servo is OFF then ON, (b) Normalized histogram of the combined laser power evolution according to time when the servo is OFF (black) then ON (red). (c) Power spectral density of the 7-fiber laser array when the NN servo is OFF (black) and ON (red) and their moving average (green and blue traces, respectively).

When TANN computed the NN in the phase correction loop for setting a non-uniform phase map, the excellent operation of the system was preserved. Few examples of some specific phase charts, most of which can be easily recognized by the naked eye, are given in Figure 9. The desired phase map for the beam array can be any arbitrary phase state. It could be changed on-demand in real-time during the laser system operation. Figure 10a reports a sequence of repeated variation in the desired target. The vertical scale denotes the errors in the individual beams' phase with respect to their steady state values corresponding to the desired state. The parameter presents an intensity correlation between the scattered pattern at the time considered and the one at the end of each cycle. Periodically the demanded phase chart was changed, and there was a sudden drop of this parameter. Each time, the system quickly restored a value close to the maximum achievable. This means the system repeatedly achieved a fast and stable setting to the new requested phase relationships. Figure 10b presents the statistical data of experimental convergence to 1000 arbitrary target phase maps, on a very short time scale. This graph shows that, regardless of the target phases, the TANN phase control system set the fiber laser output of the desired phases within about six rounds of correction, i.e., here within 550 µs.

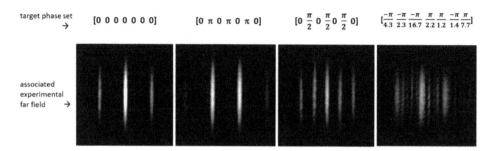

Figure 9. Examples of experimental far field patterns of phase-locked fiber laser output and their associated target phase sets.

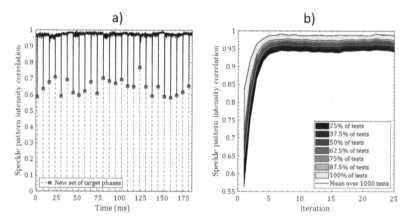

Figure 10. Experimental sequence of periodic target phase changes showing the evolution of the speckle pattern intensity correlation. Vertical scale denotes errors in the individual beams' phase with respect to their steady state values corresponding to the desired state. (**a**) Red dotted lines mark the times of target phase changes, (**b**) same experimental sequence of data folded in a single cycle, highlighting the dynamics toward a steady state phase profile for 1000 abrupt phase changes. One iteration of the phase correction loop took 92 µs.

6. Conclusions

We have reported an improved version of a phase-locking technique for a laser beam array based on neural network and quasi reinforcement learning that offers a quick on-demand change of the transverse phase distribution in the array. The NN is included in a feedback loop and computes the phase correction from data measured in a scattered pattern of the output. Instead of learning the NN for a given target, as previously studied, the original idea presented here is in the learning of a preliminary network TANN that will compute the NN parameters suited to the desired phase map. The calculation by TANN is on an order of magnitude faster than the NN training duration. Thus, the NN quickly accommodates any change of the desired phase set, so that the new architecture forms an actual adaptive phase-locking system. We first analyzed the proposed approach by simulation of an array of 2 to 20 beams. The training time of TANN was short, requiring approximately 5 min for 20 beams. The phasing accuracy was high with the NN computed by TANN, and the dynamics for phase-locking were fast, needing only a few (three iterations on average for a seven-beam array) phase error correction steps, regardless of the target phase set. The impact on the performances concerning sparsity in the sampling of the scattered pattern which was employed in the phase-sensing module was analyzed. A rule of thumb was derived for the lowest number of measurements in order to obtain a sufficiently high phasing accuracy. The technique can be applied to any form of geometry of the near field array including 1D, 2D, triangular or square lattices, rings, etc.

In the second step, we implemented the technique on a 7-channel fiber laser delivering multi-watt linearly polarized laser radiation at 1064 nm in a 1D-beam array. This experiment, with double-stage fiber amplifiers, demonstrated the efficiency of the quasi-reinforcement learning approach to set and lock the array output on a requested target phase set. This represents, to the best of our knowledge, the first time that a real laser beam array, with many independent and long amplifying arms, was phase-locked using an NN approach. The phase-lock loop featured a phasing accuracy close to $\lambda/20$ RMS and a measured bandwidth above 1 kHz. We presented the adaptive behavior of the system with respect to the target choice and analyzed its dynamics. The time response to a new request was measured at approximately 550 µs, in the non-optimized configuration. It is sufficiently fast, for example, to compensate for first order perturbations of the atmosphere in cases where the device would be connected to an appropriate sensor.

Author Contributions: Formal analysis: M.S., G.M. and P.A.; Investigation: M.S.; Project administration: V.K.; Software: G.M.; Supervision: P.A. and A.D.-B.; Validation: A.B. (Alexandre Boju).; Visualization: A.B. (Alexandre Boju) and A.B. (Alain Barthelemy).; Writing—original draft: A.B. (Alain Barthelemy) and V.K.; Writing—review and editing.: A.D.-B. All authors have read and agreed to the published version of the manuscript.

Funding: This research was funded by the Agence Nationale de la Recherche (ANR-10-LABX-0074-01) and CILAS Company (Ariane Group) under grant n °2016/0425.

Data Availability Statement: Not applicable.

Conflicts of Interest: The authors declare no conflict of interest. The funders had no role in the design of the study; in the collection, analyses, or interpretation of data; in the writing of the manuscript, nor in the decision to publish the results.

References

1. Ma, P.; Chang, H.; Ma, Y.; Su, R.; Qi, Y.; Wu, J.; Li, C.; Long, J.; Lai, W.; Chang, Q.; et al. 27.1 kW coherent beam combining system based on a seven-channel fiber amplifier array. *Opt. Laser Tech.* **2021**, *140*, 107016. [CrossRef]
2. Weyrauch, T.; Vorontsov, M.; Mangano, J.; Ovchinnikov, V.; Bricker, D.; Polnau, E.; Rostov, A. Deep turbulence effects mitigation with coherent combining of 21 laser beams over 7 km. *Opt. Lett.* **2016**, *41*, 840–842. [CrossRef]
3. Yang, X.; Huang, G.; Li, F.; Li, X.; Li, B.; Geng, C.; Li, X. Continuous Tracking and Pointing of Coherent Beam Combining System via Target-in-the-Loop Concept. *IEEE Phot. Tech. Lett.* **2021**, *33*, 1119–1122. [CrossRef]
4. Hou, T.; Dong, Z.; Tao, R.; Ma, Y.; Zhou, P.; Liu, Z. Spatially-distributed orbital angular momentum beam array generation based on greedy algorithms and coherent combining technology. *Opt. Express* **2018**, *26*, 14945–14958. [CrossRef] [PubMed]
5. Veinhard, M.; Bellanger, S.; Daniault, L.; Fsaifes, I.; Bourderionnet, J.; Larat, C.; Lallier, E.; Brignon, A.; Chanteloup, J.C. Orbital angular momentum beams generation from 61 channels coherent beam combining femtosecond digital laser. *Opt. Lett.* **2021**, *46*, 25–28. [CrossRef] [PubMed]
6. Bourderionnet, J.; Bellanger, C.; Primot, J.; Brignon, A. Collective coherent phase combining of 64 fibers. *Opt. Express* **2011**, *19*, 17053–17058. [CrossRef]
7. Shay, T.M.; Benham, V.; Baker, J.T.; Ward, B.; Sanchez, A.D.; Culpepper, M.A.; Pilkington, D.; Spring, J.; Nelson, D.J.; Lu, C.A. First experimental demonstration of self-synchronous phase locking of an optical array. *Opt. Express* **2006**, *14*, 12015–12021. [CrossRef] [PubMed]
8. Vorontsov, M.A.; Carhart, G.W.; Ricklin, J.C. Adaptive phase-distortion correction based on parallel gradient-descent optimization. *Opt. Lett.* **1997**, *22*, 907–909.
9. Vorontsov, M.A.; Sivokon, V. Stochastic parallel-gradient-descent technique for high-resolution wave-front phase-distortion correction. *J. Opt. Soc. Am. A* **1998**, *15*, 2745–2758. [CrossRef]
10. Yu, C.; Augst, S.; Redmond, S.; Goldizen, K.C.; Murphy, D.; Sanchez, A.; Fan, T. Coherent combining of a 4 kw, eight-element fiber amplifier array. *Opt. Lett.* **2011**, *36*, 2686–2688. [CrossRef] [PubMed]
11. Zhou, P.; Liu, Z.; Wang, X.; Ma, Y.; Ma, H.; Xu, X.; Guo, S. Coherent beam combining of fiber amplifiers using stochastic parallel gradient descent algorithm and its application. *IEEE J. Sel. Top. Quantum Electron* **2009**, *15*, 248–256. [CrossRef]
12. Kabeya, D.; Kermene, V.; Fabert, M.; Benoist, J.; Saucourt, J.; Desfarges-Berthelemot, A.; Barthélémy, A. Efficient phase-locking of 37 fiber amplifiers by phase-intensity mapping in an optimization loop. *Opt. Express* **2017**, *25*, 13816–13821. [CrossRef]

13. Boju, A.; Maulion, G.; Saucourt, J.; Leval, J.; Ledortz, J.; Koudoro, A.; Berthomier, J.-M.; Naiim-Habib, M.; Armand, P.; Kermene, V.; et al. Small footprint phase locking system for a large tiled aperture laser array. *Opt. Express* **2021**, *29*, 11445–11452. [CrossRef] [PubMed]
14. Saucourt, J.; Armand, P.; Kermène, V.; Desfarges-Berthelemot, A.; Barthélémy, A. Random Scattering and Alternating Projection Optimization for Active Phase Control of a Laser Beam Array. *IEEE Photonics J.* **2019**, *11*, 1503909. [CrossRef]
15. Tunnermann, H.; Shirakawa, A. Deep reinforcement learning for coherent beam combining applications. *Opt. Express* **2019**, *27*, 24223–24230. [CrossRef] [PubMed]
16. Hou, T.; An, Y.; Chang, Q.; Ma, P.; Li, J.; Zhi, D.; Huang, L.; Su, R.; Wu, J.; Ma, Y.; et al. Deep Learning-based phase control method for coherent beam combining systems. *High Power Laser Sci. Eng.* **2019**, *7*, e59. [CrossRef]
17. Chang, Q.; An, Y.; Hou, T.; Su, R.; Ma, P.; Zhou, P. Phase-locking System in Fiber Laser Array through Deep Learning with Diffusers, Paper M4A.96. In Proceedings of the Asia Communications and Photonics Conference, Beijing, China, 24–27 October 2020.
18. Hou, T.; An, Y.; Chang, Q.; Ma, P.; Li, J.; Huang, L.; Zhi, D.; Wu, J.; Su, R.; Ma, Y.; et al. Deep-learning-assisted, two-stage phase control method for high-power mode-programmable orbital angular momentum beam generation. *Photonics Res.* **2020**, *8*, 715–722. [CrossRef]
19. Tünnermann, H.; Shirakawa, A. Deep reinforcement learning for tiled aperture beam combining in a simulated environment. *JPhys Photonics* **2021**, *3*, 015004. [CrossRef]
20. Wang, D.; Du, Q.; Zhou, T.; Li, D.; Wilcox, R. Stabilization of the 81-channel coherent beam combination using machine learning. *Opt. Express* **2021**, *29*, 5694–5709. [CrossRef] [PubMed]
21. Zhang, X.; Li, P.; Zhu, Y.; Li, C.; Yao, C.; Wang, L.; Dong, X.; Li, S. Coherent beam combination based on Q-learning algorithm. *Opt. Comm.* **2021**, *490*, 126930. [CrossRef]
22. Shpakovych, M.; Maulion, G.; Kermene, V.; Boju, A.; Armand, P.; Desfarges-Berthelemot, A.; Barthélemy, A. Experimental phase control of a 100 laser beam array with quasi-reinforcement learning of a neural network in an error reduction loop. *Opt. Express* **2021**, *29*, 12307–12318. [CrossRef]
23. Nabors, C. Effects of phase errors on coherent emitter arrays. *Appl. Optics.* **1994**, *33*, 2284–2289. [CrossRef]
24. PPopoff, S.; Lerosey, M.G.; Carminati, R.; Fink, M.; Boccara, C.; Gigan, S. Measuring the transmission matrix in optics: An approach to the study and control of light propagation in disordered media. *Phys. Rev. Lett.* **2010**, *104*, 100601. [CrossRef]
25. Drémeau, A.; Liutkus, A.; Martina, D.; Katz, O.; Schülke, C.; Krzakala, F.; Gigan, S.; Daudet, L. Reference-less measurement of the transmission matrix of a highly scattering material using a DMD and phase retrieval techniques. *Opt. Express* **2015**, *23*, 11898–11911. [CrossRef]

Article

Adaptive Detection of Wave Aberrations Based on the Multichannel Filter

Pavel A. Khorin [1], Alexey P. Porfirev [2] and Svetlana N. Khonina [1,2,*]

1 Samara National Research University, 443086 Samara, Russia; paul.95.de@gmail.com
2 Image Processing Systems Institute of RAS—Branch of the FSRC "Crystallography and Photonics" RAS, 443001 Samara, Russia; porfirev.alexey@ipsiras.ru
* Correspondence: khonina@ipsiras.ru

Abstract: An adaptive method for determining the type and magnitude of aberration in a wide range is proposed on the basis of an optical processing of the analyzed wavefront using a multichannel filter matched to the adjustable Zernike phase functions. The approach is based on an adaptive (or step-by-step) compensation of wavefront aberrations based on a dynamically tunable multichannel filter implemented on a spatial light modulator. For adaptive filter adjustment, a set of criteria is proposed that takes into account not only the magnitude of the correlation peak, but also the maximum intensity, compactness, and orientation of the distribution in each diffraction order. The experimental results have shown the efficiency of the proposed approach for detecting wavefront aberrations in a wide range (from 0.1λ to λ).

Keywords: wavefront aberrations; adaptive method; Zernike functions; wavefront sensor; multichannel diffractive optical element

1. Introduction

The problem of measuring and correcting wavefront aberrations is often encountered in optics, for example, in the design of ground-based telescopes, in optical communication systems, in industrial laser technology, and in medicine [1–12]. Usually, the measurement of wavefront distortions is performed in order to compensate them, in particular, with adaptive or active optics [13–18]. The major causes of wavefront aberrations are turbulence of the atmosphere, imperfect shapes of the optical elements of the system, errors in the alignment of the system, etc.

It is known that weak wavefront aberrations (level $\leq 0.4\lambda$) are well detected using spatial filters matched to the basis of Zernike functions [19–27] including multichannel diffractive optical elements (DOEs) [21,25,27]. However, with an increase in aberration level, the linear approximation of the wavefront by Zernike functions becomes unacceptable [27]. This is explained by the fact that the contribution of the second and subsequent nonlinear terms of the wavefront expansion to the Taylor series becomes more significant, which leads to the detection of false aberrations.

With high aberrations (level > 0.4λ), when a significant blurring of the focal spot occurs, it makes sense to use methods focused on analyzing the intensity distribution pattern formed by an aberrated optical system in one or several planes. To determine the wavefront in this case, iterative [28–32] and optimization algorithms [10,33] are used, including those with the use of neural networks [34–39]. In turn, these approaches demonstrate significant errors for small aberrations, when the point spread function (PSF) is close to the Airy picture of an ideal system [27].

Thus, different methods work at different levels of aberrations, and in order to apply them, it is desirable to determine this level (or magnitude). One of the solutions is the use of additional optical and digital processing, for example, based on a dynamically tunable spatial light modulator (SLM). Previously, we studied the stability of the wavefront expansion coefficients on the basis of Zernike polynomials during the field propagation in free space [40], the application limits of spatial filters matched with the basis of Zernike functions [41–43], and the possibility of scaling aberrations levels for testing optical systems [44].

In this paper, to determine the type and magnitude (or level) of aberrations in the investigated wavefront (WF), the application of an adaptive method based on the use of a multichannel filter matched with adjustable Zernike phase functions (Zernike polynomials correspond to the phase of the considered functions) is proposed. In this case, instead of the optical expansion of the field on the basis of the Zernike functions which was realized earlier [21,25,27], we actually perform multichannel aberration compensation based on the Zernike polynomials. The novelty of our study lies in the combination of an adaptive approach and matched filtering based on a multichannel diffractive optical element. Note that phase compensation can occur in each channel in accordance with different types of aberrations, as well as for the same aberration type but with different magnitude (or level). The method is based on a step-by-step compensation of wavefront aberrations based on a dynamically tunable multichannel filter implemented on a spatial light modulator. A set of criteria for adaptive filter tuning is proposed, taking into account not only the correlation peak presence, but also the maximum intensity, compactness, and orientation of the PSF distribution in each diffraction order. The experimental results have shown the efficiency of the proposed approach for detecting wavefront aberrations in a wide range (from 0.1λ to λ).

2. Materials and Methods

2.1. Theoretical Foundations

Wavefront aberrations are usually described in terms of Zernike functions $Z_{nm}(r, \varphi)$ [45] in the following way:

$$W(r, \varphi) = \exp\left[i\kappa\alpha \sum_{n=0}^{N_{max}} \sum_{m=0}^{n} C_{nm} Z_{nm}(r, \varphi)\right], \tag{1}$$

where $\kappa = 2\pi/\lambda$ is the wave number, λ is the radiation wavelength, α is the parameter corresponding to the level of aberration in wavelengths (in this paper it takes values from zero to wavelength λ), and coefficients C_{nm} are normalized.

The PSF for an aberrated wavefront can be calculated using the Fourier transform:

$$F(u, v) = \Im\{W(x, y)\} = \int_{-\infty}^{\infty} \int_{-\infty}^{\infty} W(x, y) \exp\left[-i\frac{2\pi}{\lambda f}(ux + vy)\right] dx\, dy, \tag{2}$$

where f is the focal length of the lens.

The optical wavefront analyzer considered in this work is based on the addition of a lens with a diffractive multichannel filter [21,25,27] of the following form:

$$\tau(x, y) = \sum_{p=0}^{P} \sum_{q=0}^{Q} \Psi_{pq}^*(x, y) \exp\left[i(a_{pq}x + b_{pq}y)\right], \tag{3}$$

where $\Psi_{pq}(x, y)$ are the functions matched to diffraction orders (DOs) with indices (p, q), the position of which in the focal plane is determined by the spatial frequencies a_{pq} and b_{pq}.

As a rule, the Zernike polynomials are used as $\Psi_{pq}(x, y)$ [7,21,25,27] to analyze a wavefront. However, correct detection in this case is possible only at low levels of aberrations. In [27], it was shown that at an aberration level of more than 0.4λ, the use of such a filter leads to false detection.

In this paper, the use of a filter matched with the Zernike phase functions (Zernike polynomials correspond to the phase of the considered functions) is proposed:

$$\Psi_{pq}(x,y) = \exp[i\kappa\alpha_0 Z_{pq}(x,y)], \quad (4)$$

where $Z_{pq}(x,y)$ are the standard Zernike functions of double indices (p, q) [45]. The correspondence of Zernike functions to various types of aberrations is given in Table 1.

Table 1. Correspondence of Zernike functions of double indices (n, m) to the aberration type.

(n, m)	(0, 0)	(1, 1)	(2, 2)	(2, 0)	(3, 3)	(3, 1)	(4, 4)	(4, 2)	(4, 0)
$Z_{nm}(x,y)$	1	$2r\cos(\varphi)$	$\sqrt{6}r^2\cos(2\varphi)$	$\sqrt{3}(2r^2-1)$	$2\sqrt{2}r^3\cos(3\varphi)$	$2\sqrt{2}(3r^3-2r)\cos(\varphi)$	$\sqrt{10}r^4\cos(4\varphi)$	$\sqrt{10}(4r^4-3r^2)\cos(2\varphi)$	$\sqrt{5}(6r^4-6r^2+1)$
Aberration type	Ideal	Tilt	Astigmatism	Defocus	Coma (Trefoil)	Pure coma	Quadrofoil	2nd order astigmatism	Spherical

2.2. Scheme for Adaptive Aberration Compensation

We consider in detail several wave aberrations of different magnitudes and show the scheme for the adaptive method to compensate these aberrations.

We divide the aberrations into even type (n is even) and odd type (n is odd) and consider them separately. Among the even aberrations, one can distinguish axis-symmetric ones (m is equal to 0). Even aberrations such as (2, 0), (4, 0), ..., $(n, 0)$ have the wavefront phase and corresponding PSF axis-symmetric distribution regardless of the level of aberration α (see Table 2).

Table 2. Simulation results for radial wave aberration of type $(n, m) = (2, 0)$ and $(n, m) = (4, 0)$.

Aberration Type	α	0.10λ	0.25λ	0.50λ	0.75λ	1.00λ
Defocus $(n, m) = (2, 0)$	WF phase					
	PSF					
Spherical $(n, m) = (4, 0)$	WF phase					
	PSF					

The adaptive method is described as the aberration compensation by complex conjugate functions (4) with different levels of α_0. In other words, we calculate the compensated wavefront of the form

$$\Delta W(x,y) = W(x,y)\Psi^*_{pq}(x,y) = \exp[i\kappa\alpha Z_{nm}(x,y)]\exp[-i\kappa\alpha_0 Z_{pq}(x,y)] \quad (5)$$

and the corresponding PSF of the form

$$\Delta F(u,v) = \Im\{\Delta W(x,y)\} = \int_{-\infty}^{\infty}\int_{-\infty}^{\infty} \Delta W(x,y)\exp\left[-i\frac{2\pi}{\lambda f}(ux+vy)\right]dxdy, \quad (6)$$

where $W(x,y)$ is the analyzed wavefront with aberration (n, m) and level α, $\Psi^*_{pq}(x,y)$ is the complex conjugate function (4) with a variable level α_0, $\Delta W(x,y)$ is the resulting wavefront, and $\Delta F(x,y)$ is the corresponding PSF.

With the successful compensation of the specified wave aberration, the phase of the resulting wavefront should be equal to a constant, and the PSF should correspond to the Airy picture (PSF for the plane WF).

The results collected in Figure 1 show the compensating process for the spherical aberration $(n, m) = (2, 0)$ with a level $\alpha = 0.5\lambda$ by the set of $\Psi^*_{pq}(x,y) = \Psi^*_{20}(x,y)$ with different levels of $\alpha_0 = \{0.1\lambda; 0.25\lambda; 0.5\lambda; 0.75\lambda\}$. Thus, we construct the resulting wavefront defined by Equation (5) and the corresponding PSF defined by Equation (6).

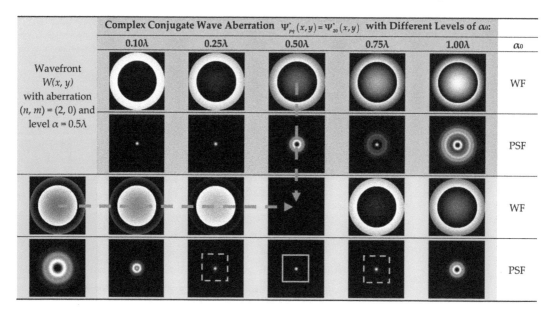

Figure 1. Illustration of the compensating process for wave aberration $W(x, y)$ of type $(n, m) = (2, 0)$ with $\alpha = 0.5\lambda$ level (green highlights) by complex conjugate wave aberration $\Psi^*_{pq}(x,y) = \Psi^*_{20}(x,y)$ with different levels of α_0 (orange highlights). Results of compensation (blue highlights) are placed at the intersection of corresponding rows and columns.

Figure 1 shows in detail the effect of the compensating wavefront $W(x, y)$ with aberration type $(n, m) = (2, 0)$ and level $\alpha = 0.5\lambda$ by $\Psi^*_{pq}(x,y) = \Psi^*_{20}(x,y)$ with different levels of α_0. The compensated phase (equal to 0) is observed at the intersection of column 4 and row 3, and the corresponding PSF (the Airy picture $D(u,v)$) is observed at the intersection of column 4 and row 4:

$$\Delta W(x,y) = W(x,y)\Psi^*_{pq}(x,y) = \exp[i2\pi 0.5 Z_{20}(x,y)]\exp[-i2\pi 0.5 Z_{20}(x,y)] = 1, \quad (7)$$

$$\Delta F(u,v) = \Im\{\Delta W(x,y)\} = \int_{-\infty}^{\infty}\int_{-\infty}^{\infty}\exp\left[-i\frac{2\pi}{\lambda f}(ux+vy)\right]dxdy = D(u,v). \quad (8)$$

In addition, it can be seen from Figure 1 that the PSFs observed in row 4, columns 3 and 5, have the same intensity. This is because the phase of the resulting wavefront for these two cases is the same up to sign:

$$\Delta W(x,y) = \exp[i2\pi 0.5 Z_{20}(x,y)] \exp[-i2\pi 0.25 Z_{20}(x,y)] = \exp[i2\pi 0.25 Z_{20}], \quad (9)$$

$$\Delta W(x,y) = \exp[i2\pi 0.5 Z_{20}(x,y)] \exp[-i2\pi 0.75 Z_{20}(x,y)] = \exp[-i2\pi 0.25 Z_{20}]. \quad (10)$$

A change in the sign of the aberration coefficient to the opposite causes a symmetric transformation of the PSF image about the horizontal axis for negative m, since $-\sin(m\varphi) = \sin(m\varphi + \pi)$, and about the vertical axis for positive m, since $-\cos(m\varphi) = \cos(m\varphi + \pi)$. For even aberrations, this property is not so noticeable due to their symmetry. Therefore, PSFs for even wave aberrations that differ only in the sign of the weighting coefficient have the same intensity distribution in the focal plane.

Even aberrations with $m \neq 0$ have phase images with m symmetry axes and PSFs with $2m$ symmetry axes. It should be noted that the OX and OY axes are always the PSF symmetry axes (see Table 3).

Table 3. Results of modeling an even wave aberration of the type $(n, m) = (2, 2)$ and $(n, m) = (4, 4)$.

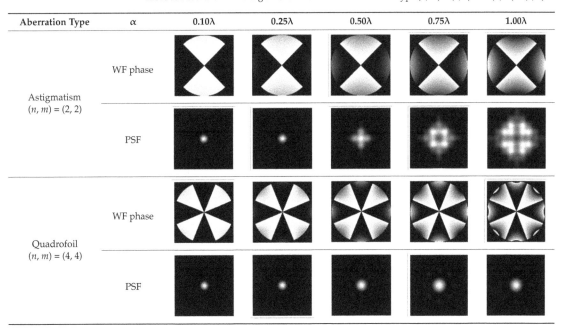

Analogical results of the compensating process for the astigmatic aberration $(n, m) = (2, 2)$ with a level $\alpha = 0.5\lambda$ by the set of $\Psi^*_{pq}(x,y) = \Psi^*_{22}(x,y)$ with different levels of $\alpha_0 = \{0.1\lambda; 0.25\lambda; 0.5\lambda; 0.75\lambda\}$ are shown in Figure 2. The situation is similar to that of Figure 1, and cases at $\alpha_0 = \{0.25\lambda; 0.75\lambda\}$ correspond to identical PSFs.

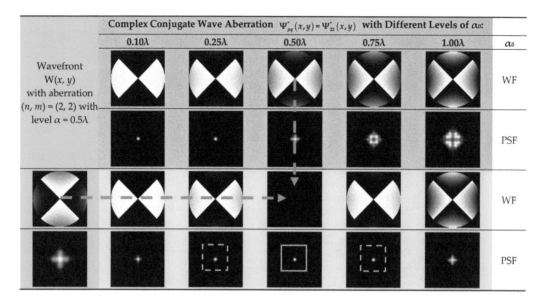

Figure 2. Illustration of the compensating process for wave aberration $W(x, y)$ of type $(n, m) = (2, 2)$ with $\alpha = 0.5\lambda$ level by complex conjugate wave aberration $\Psi^*_{pq}(x,y) = \Psi^*_{22}(x,y)$ with different levels of α_0 (description for pictures are the same as for Figure 1).

Odd-type aberrations such as $(n, m) = (3, 1)$ and $(n, m) = (3, 3)$ have images of the wavefront phase with m symmetry axes and the corresponding PSFs with $2m$ symmetry axes (see Table 4).

Table 4. Modeling results for odd wave aberration of type $(n, m) = (3,1)$ and $(n, m) = (3, 3)$.

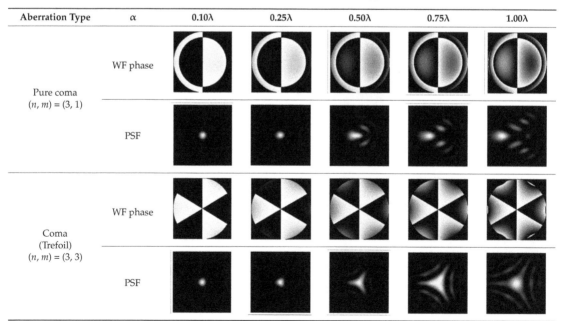

The results of the compensating process for the coma-type aberration $(n, m) = (3, 1)$ with a level $\alpha = 0.5\lambda$ by the set of $\Psi^*_{pq}(x,y) = \Psi^*_{31}(x,y)$ with different levels of $\alpha_0 = \{0.1\lambda; 0.25\lambda; 0.5\lambda; 0.75\lambda\}$ are shown in Figure 3. The situation is different from even aberrations.

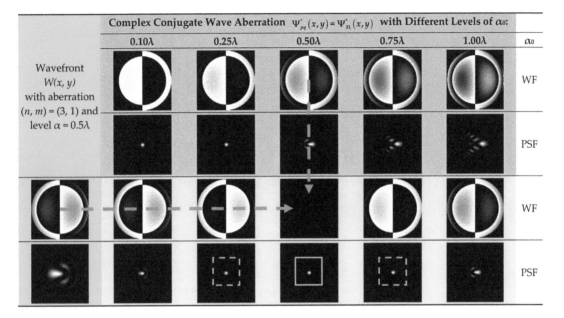

Figure 3. Illustration of the compensating process for wave aberration $W(x, y)$ of type $(n, m) = (3, 1)$ with $\alpha = 0.5\lambda$ level (green highlights) by complex conjugate wave aberration $\Psi^*_{pq}(x,y) = \Psi^*_{31}(x,y)$ with different levels of α_0 (orange highlights). Results of compensation (blue highlights) are placed at the intersection of corresponding rows and columns.

It can be seen from Figure 3 that in row 4, columns 3 and 5, PSFs are rotated by 180 degrees. This is because the phase of the resulting wavefront for these two cases is the same up to sign:

$$\Delta W(x,y) = \exp[i2\pi 0.5 Z_{31}(x,y)] \exp[-i2\pi 0.25 Z_{31}(x,y)] = \exp[i2\pi 0.25 Z_{31}], \quad (11)$$

$$\Delta W(x,y) = \exp[i2\pi 0.5 Z_{31}(x,y)] \exp[-i2\pi 0.75 Z_{31}(x,y)] = \exp[-i2\pi 0.25 Z_{31}]. \quad (12)$$

Changing the sign of the aberration coefficient to the opposite one causes a symmetric transformation of the PSF image about the horizontal axis. As described earlier for positive m, such a symmetric transformation must be performed with respect to the vertical axis, since $-\cos(m\varphi) = \cos(m\varphi + \pi)$. For odd-type aberrations, this property is evidentially observed in the PSF intensity distribution.

Thus, the introduced empirical criterion for odd-type aberrations is based on the described fundamental properties of the PSF of wave aberrations.

The following combined criteria for correct detection can be used:

1. **Formation of a pronounced/evident correlation peak** in the focal plane of the filter in the DO with indices (p, q).
2. **Change in the orientation of the PSF intensity distribution** in the adjacent DOs of the "level and type" (LT) filter within one row (for odd-symmetry aberrations).
3. **Presence of the same PSF intensity distributions** in the adjacent DOs of the LT-filter within one row (for even-symmetry aberrations).

Based on the analysis of the intensity distribution in the focal plane, taking into account the proposed criteria, it is possible to determine the range of the analyzed aberration level.

Next, we propose using a multichannel filter to implement the processing of the analyzed wavefront simultaneously by several types of aberrations taken into account.

2.3. Multichannel Filter Employment

To determine the type and magnitude of aberration in the analyzed wavefront, we propose combining the adaptive method with the use of a multichannel filter matched to Zernike's adjustable phase functions. Thus, the principle of operation of our method lies in the parallel application of the adaptive approach by a multichannel diffractive element employment. In this case, a wide range of detectable aberration values and resistance to vibrations are provided.

To realize this approach, the filter should be matched with a set of functions defined by Equation (4) with different levels of α_k. We can design a filter tuned to one specific wave aberration with a value ranging from α_1 to $\alpha_{K\max}$.

In order to detect another type of wave aberrations, it is necessary to divide each channel into $P_{\max}Q_{\max}$ additional channels. So, different wave aberrations with different α_k should be encoded in different diffraction orders. The transmission function of such an LT-filter will be as follows:

$$\tau_{LT}(x,y) = \sum_{p=0}^{P_{\max}} \sum_{q=0}^{Q_{\max}} \sum_{k=1}^{K_{\max}} \exp\left[-i\kappa\alpha_k Z_{pq}(x,y)\right] \exp\left[i\left(a_{kpq}x + b_{kpq}y\right)\right]. \quad (13)$$

It should be noted that the accuracy of the detected (and compensated) aberration value directly depends on the number of channels $P_{\max}Q_{\max}K_{\max}$; however, the number of filter channels is limited by technological possibility. Since the filter defined by Equation (13) can be implemented using an SLM, the number of channels can be reduced taking into account a certain time delay for rebuilding the filter to different sets of channels. Moreover, in the presence of an additional adapting device, it is possible to organize an iterative process of step-by-step compensation of aberrations [30–32].

The formation of a correlation peak in one of the DOs can be considered as a criterion for successful compensation. Additional criteria for adaptive filter tuning are proposed in Section 3.

A special case of the LT-filter defined by Equation (13) is the "type" T-filter tuned for different types of aberrations with the same level α_0:

$$\tau_T(x,y) = \sum_{p=0}^{P_{\max}} \sum_{q=0}^{Q_{\max}} \exp\left[-i\kappa\alpha_0 Z_{pq}(x,y)\right] \exp\left[i\left(a_{pq}x + b_{pq}y\right)\right]. \quad (14)$$

If the aberration type is determined and we need to more accurately determine its level, then it makes sense to use another particular version of the filter defined by Equation (13), namely the "level" L-filter tuned to only one aberration type (p_0, q_0) with different levels of α_k:

$$\tau_L(x,y) = \sum_{k=1}^{K_{\max}} \exp\left[-i\kappa\alpha_k Z_{p_0 q_0}(x,y)\right] \exp[i(a_k x + b_k y)]. \quad (15)$$

To analyze other aberration types, the filter can be quickly rebuilt using an SLM.

We have designed several variants of multichannel filters defined by Equations (13)–(15). Figure 4 demonstrates three examples of these filters: *T-filter* matched with different types of aberrations of the same level (Equation (14), Figure 4a), *LT-filter* matched with different types and levels of aberrations (Equation (13), Figure 4b), and *L-filter* matched with one aberration type of different levels (Equation (15), Figure 4c). Each of the options is suitable for a specific task. It makes sense to use a hybrid LT-filter for preliminary analysis and a filter consistent with different levels for refinement.

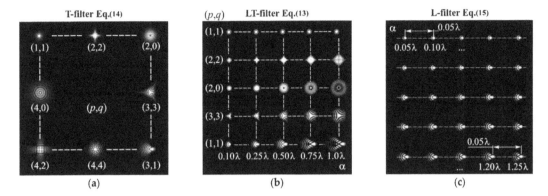

Figure 4. Examples of intensity patterns for different filters under plane wavefront illumination: (**a**) T-filter defined by Equation (14) at $\alpha_0 = 0.5$; (**b**) LT-filter defined by Equation (13), and (**c**) L-filter defined by Equation (15) at $(p_0, q_0) = (3, 1)$.

With the help of dynamically tunable filters implemented on an SLM, it becomes possible to specify the level of aberrations of the analyzed WF or to take the average value from the already detected range. Note that the accuracy of this range directly depends on the noise level of the recording device.

2.4. Optical Setup

The optical scheme used in the experiment is shown in Figure 5.

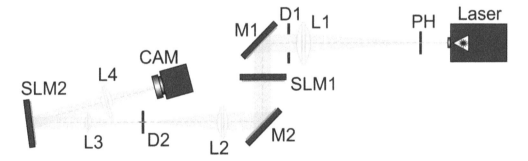

Figure 5. Schematic of the experimental setup for detection of wavefront aberrations and their superpositions using a diffractive multichannel filter. Laser is a solid-state laser ($\lambda = 532$ nm); PH is a pinhole (hole size of 40 μm); L1, L2, L3, and L4 are spherical lenses ($f_1 = 350$ mm, $f_2 = 300$ mm, $f_3 = 200$ mm, and $f_4 = 250$ mm); SLM1 is a transmissive spatial light modulator (HOLOEYE LC 2012); SLM2 is a reflective spatial light modulator (HOLOEYE PLUTO VIS); D1 and D2 are circular apertures; M1 and M2 are mirrors; CAM is a ToupCam UCMOS08000KPB video camera.

The experimental setup for detecting wavefront aberrations using a diffraction multi-channel filter consists of the following elements: a solid-state laser ($\lambda = 532$ nm), a pinhole (hole size of 40 μm), spherical lenses ($f_1 = 350$ mm, $f_2 = 300$ mm, $f_3 = 200$ mm, and $f_4 = 250$ mm), the transmissive SLM (HOLOEYE LC 2012), the reflective SLM (HOLOEYE PLUTO VIS), circular apertures, mirrors, and the video camera ToupCam UCMOS08000KPB.

The output laser radiation from a solid-state laser ($\lambda = 532$ nm) was collimated using a system consisting of a pinhole (PH) with a hole diameter of 40 μm and a spherical lens (L_1) ($f_1 = 350$ mm). A circular diaphragm D1 was used to separate a central light spot from surrounding light and dark rings occurring during diffraction at the pinhole. Then, the laser beam expanded and reflected by mirror M1 passed through a HOLOEYE LC 2012 transmissive spatial light modulator SLM1 with a 1024 × 768 pixel resolution and a pixel size of 36 μm, which was used to form a wavefront with a required combination of aberrations. Lenses L2 (f_2 = 300 mm) and L3 (f_3 = 200 mm) and diaphragm D2 were used to spatially split the aberration-distorted laser beam formed by the first modulator and the unmodulated zero-order transmitted laser beam.

A mirror M2 was used to direct the formed laser beam to the display of the second modulator. A HOLOEYE PLUTO VIS reflective spatial light modulator SLM2 with a 1920 × 1080 pixel resolution and a pixel size of 8 μm was used to implement a phase mask of a diffractive multichannel filter, which served to decompose the studied light field in terms of the Zernike polynomial basis. The laser beam reflected from this SLM was directed to lens L4 (f_4 = 250 mm), which focused it on the matrix of a ToupCam UCMOS08000KPB camera CAM with a 3264 × 2448 resolution and a pixel size of 1.67 μm.

3. Results

3.1. Detection of One Type of Wave Aberration

As a rule, when using matched filters, the appearance of a correlation peak is a criterion for detecting a matched signal [46–50]. Note that the presence of a correlation peak is determined by a nonzero intensity at the center of the DO. However, when analyzing phase distributions, this approach is correct only for small phase variations. In particular, it was shown in [27] that when the aberration level is more than 0.4λ, the use of filters matched with the basis of Zernike functions leads to the appearance of many wrong correlation peaks and false detection.

The adaptive approach considered in Section 2.2 is free from this limitation, but for its application, it is necessary to select suitable compensating functions in accordance with the developed criteria. To speed up the selection process, we use multichannel filters to analyze a wavefront simultaneously by several types of compensating functions.

In this section, we demonstrate the use of the multichannel filters described in Section 2.3 for test examples when aberrations of certain types with a given level are introduced into the wavefront.

Figure 6 and Table 5 show the results of a numerical and experimental analysis of the WF distorted by a coma-type aberration $(n, m) = (3, 1)$ with a weight (level) $\alpha = 0.5\lambda$, using various diffractive multichannel filters (the intensity patterns in the focal plane are shown). DOs with correlation peaks are framed by a solid line, and DOs with changed PSF orientation are framed by a dashed line. The selected (framed) orders correspond to the criteria formulated in Section 2.2. A good agreement between the numerical simulation and the optical experiment is seen.

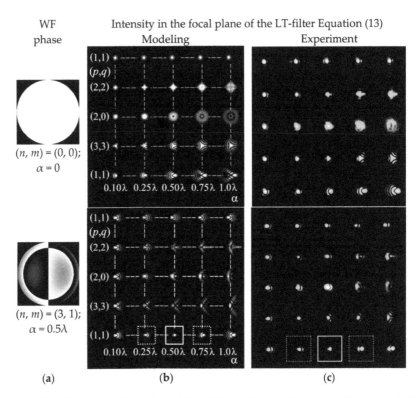

Figure 6. Detailed results of the analysis of the WF distorted by the odd-type aberration $(n, m) = (3, 1)$ with $\alpha = 0.5\lambda$, using the LT-filter: (**a**) the phase of the analyzed WF; (**b**) simulation of the filter action; (**c**) optical experiment.

The results for LT-filter show (see Figure 6) that the correlation peak located in the DO corresponding to $(p, q) = (3, 1)$ with $\alpha_k = 0.5\lambda$ and adjacent DOs contains a PSF with changed orientation (see Figure 6, positions are marked with frames). As follows from the given example, for aberrations with odd symmetry, the criterion of the correct detection can be not only the presence of the correlation peak, but also the orientation of PSF distribution in neighboring DOs. A similar result was obtained in [51,52] based on numerical simulation. In contrast to these papers, in this work, we realize experimental investigation of the proposed approach and apply it for a step-by-step compensation of a complex type of aberrations.

T-filters with different levels of α_0 also demonstrate (see Table 5) a change of PSF orientation when passing through the level $\alpha_0 = 0.5\lambda$. A correlation peak is also observed in the DO corresponding to $(p, q) = (3, 1)$ with a level equal to 0.5λ (see Table 5, third row).

The results of simulation and experiment with another test example, when the WF is distorted by an even type of aberration $(p, q) = (2, 2)$ with $\alpha = 0.5\lambda$, are presented in Figure 7. Results show that the DOs adjacent to the correlation peak have the same PSF intensity distribution. Thus, when detecting even-symmetry aberrations, in addition to the presence of a correlation peak, the criterion for correct detection can be the similarity of diffraction patterns in neighboring orders.

Table 5. Results of modeling and optical experiment of the operation of multichannel T-filters in detecting WF distorted by coma-type aberration $(n, m) = (3, 1)$ with $\alpha = 0.5\lambda$.

WF	Type of the Filter	Intensity in the Focal Plane of the T-Filter Equation (14)	
		Modeling	Experiment
$(n, m) = (0, 0)$; $\alpha = 0$	T-filter at $\alpha_0 = 0.5\lambda$	(1,1) (2,2) (2,0) / (4,0) (p,q) (3,3) / (4,2) (4,4) (3,1)	
$(n, m) = (3, 1)$; $\alpha = 0.5\lambda$	T-filter at $\alpha_0 = 0.3\lambda$	(1,1) (2,2) (2,0) / (p,q) / (4,0) (3,3) / (4,2) (4,4) (3,1)	
	T-filter at $\alpha_0 = 0.5\lambda$	(1,1) (2,2) (2,0) / (p,q) / (4,0) (3,3) / (4,2) (4,4) (3,1)	
	T-filter at $\alpha_0 = 0.7\lambda$	(1,1) (2,2) (2,0) / (p,q) / (4,0) (3,3) / (4,2) (4,4) (3,1)	

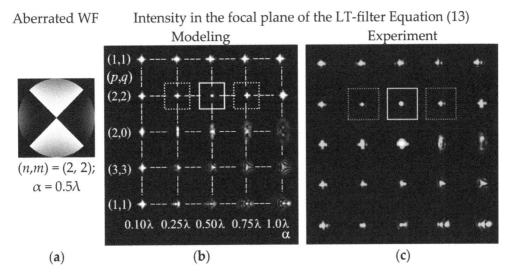

Figure 7. Detailed results of the analysis of the WF distorted by the even type of aberration $(n, m) = (2, 2)$ with $\alpha = 0.5\lambda$, using the LT-filter: (**a**) the phase of the analyzed WF; (**b**) simulation of the filter action; (**c**) optical experiment.

The results of a test example with a WF distorted by aberration $(n, m) = (3, 3)$ with an increased weight $\alpha = 0.7\lambda$ are shown in Figure 8 and Table 6.

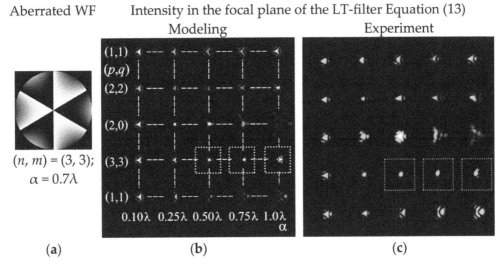

Figure 8. Results of the analysis of the WF distorted by the aberration $(n, m) = (3, 3)$ with $\alpha = 0.7\lambda$, using the LT-filter: (**a**) the phase of the analyzed WF; (**b**) simulation of the filter action; (**c**) optical experiment.

Table 6. Results of modeling and optical experiment of the operation of multichannel T-filters in detecting WF distorted by aberration $(n, m) = (3, 3)$ with $\alpha = 0.7\lambda$.

Aberrated WF	Type of the Filter	Intensity in the Focal Plane of the T-Filter Equation (14)	
		Modeling	Experiment
$(n, m) = (3, 3);$ $\alpha = 0.5\lambda$	T-filter at $\alpha_0 = 0.5\lambda$		
	T-filter at $\alpha_0 = 0.7\lambda$		
	T-filter at $\alpha_0 = 0.9\lambda$		

The increase in weight was considered to show the good performance of filters over a wide range of aberration values (if necessary, the filter can be easily reconfigured for a larger range). In the focal plane of the LT-filter (see Figure 8), the PSF changes orientation in the DOs marked with frames (which corresponds to an aberration of the $(p, q) = (3, 3)$ aberration type); however, no evident correlation peak is observed between these DOs. The absence of such a correlation peak is explained by the inconsistency of the level of aberrations recorded in the filter with the level of the analyzed wave aberration. To refine the weight of the detected aberration, it is required to use an L-filter or T-filter with different levels of aberration $(p, q) = (3, 3)$ (see Table 6).

It can be seen from Table 6 that the use of a T-filter with different levels of matched functions made it possible to refine the weight of the detected aberration $(p, q) = (3, 3)$; thus, the correlation peak is observed at $\alpha_0 = 0.7\lambda$ (see Table 6, second rows). In this case, for the levels $\alpha_0 = 0.5\lambda$ and $\alpha_0 = 0.9\lambda$ (see Table 6, first and third rows), the intensity distribution changes orientation, which is an additional sign of correct detection.

3.2. Detection of Superposition of Wave Aberrations (Complex Aberration)

In this section, we illustrate the action of the adaptive (step-by-step) method for detecting a superposition of wave aberrations, where at each l-th stage the wavefront $W_l(x,y)$ (at the first iteration it is equal to the investigated wavefront) is analyzed by the dynamically tunable SLM. In the focal plane of the filter, there are different PSF pictures. We recall that the aberration type (p, q) is associated with each string and aberration level with k-column (α_k are fitted in accordance with the combined criteria introduced above) of the diffractive multichannel filter. So, the investigated wavefront is corrected taking into account the detected wave aberration as follows:

$$W_{l+1}(x,y) = W_l(x,y)/W_\Delta(x,y) = W_l(x,y)\exp[-i\kappa\alpha_k Z_{pq}(x,y)], \quad (16)$$

where $W_\Delta(x,y) = \exp[-i\kappa\alpha_k Z_{pq}(x,y)]$.

The criterion for correctly detected aberrations is the formation of a pronounced/evident correlation peak in the focal plane of the filter in the DO at one of the l-th steps. To detect the correlation peak formation, we use digital image processing and calculate mean square error (MSE) for $W_l(x,y)$ in the region of DO (the considered region size is $S \times S$ pixels):

$$MSE[W_l(u,v), D(u,v)] = \frac{1}{S^2}\sum_{i=1}^{S}\sum_{j=1}^{S}(W_l(u_i,v_j) - D(u_i,v_j))^2. \quad (17)$$

where $D(u,v)$ is the Airy picture (PSF for plane/ideal WF).

To illustrate this approach, we consider the WF distorted by the superposition of two aberrations $0.5\lambda*(2, 2) + 0.5\lambda*(3, 3)$. At the first stage, we used an LT-filter (see Figure 9) and revealed that the DO changes orientation in the fourth row, which corresponds to type $(p, q) = (3, 3)$ with $\alpha = 0.5\lambda$, and a residual aberration of the type $(n, m) = (2, 2)$ is observed in the corresponding DO. So, at the first iteration, the wavefront aberration $(n, m) = (3, 3)$ with 0.5λ level was compensated by LT-filter. Thus, the problem was reduced by one dimension.

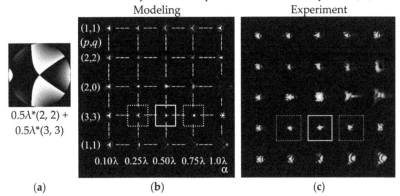

Figure 9. Results of modeling and optical experiment in detecting WF distorted by the superposition of aberrations $0.5\lambda*(2, 2) + 0.5\lambda*(3, 3)$ using the LT-filter (change in the DO orientation—dashed line): (a) the phase of the analyzed WF; (b) simulation of the filter action; (c) optical experiment.

After compensating for one detected aberration, the wavefront has residual aberration $(n, m) = (2, 2)$ with 0.5λ level. The result of the action of the same LT-filter on the obtained wavefront is presented in Figure 10. The LT-filter shows that the DOs are symmetric in the second row with respect to the correlation peak at $(p, q) = (2, 2)$ with 0.5λ level.

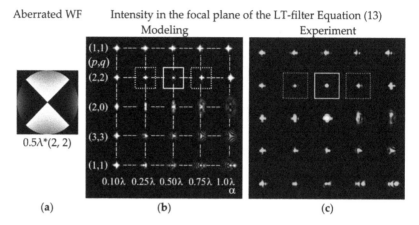

Figure 10. Results of modeling and optical experiment in detecting WF with residual aberration $(n, m) = (2, 2)$ with $\alpha = 0.5\lambda$ level using the LT-filter (change in DO orientation—dashed line; correlation peak—solid line): (**a**) the phase of the analyzed WF; (**b**) simulation of the filter action; (**c**) optical experiment.

In another test example (Figure 11), we considered WF distorted by the superposition of three aberrations $0.5\lambda*(2, 0) + 0.5\lambda*(2, 2) + 0.5\lambda*(3, 3)$. The LT-filter shows (see Figure 11) that the DO changes orientation in the fourth row, which corresponds to aberration $(p, q) = (3, 3)$ with 0.5λ level. So, at the first iteration, we compensated for one type of aberration and reduced the problem by one dimension. At the second iteration, the residual aberrations are similar to the first example (see Figure 9), which was compensated in a similar way.

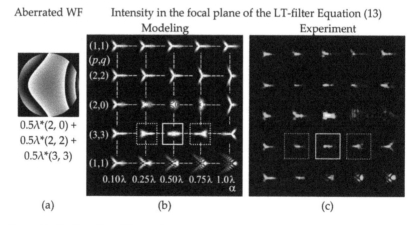

Figure 11. Results of modeling and optical experiment in detecting WF distorted by the superposition of aberrations $0.5\lambda*(2, 0) + 0.5\lambda*(2, 2) + 0.5\lambda*(3, 3)$ using the LT-filter (change in the DO orientation—dashed line): (**a**) the phase of the analyzed WF; (**b**) simulation of the filter action; (**c**) optical experiment.

To refine the weight of the desired aberration, it is possible to iteratively subtract the detected aberrations with low weights until a pronounced correlation peak is observed in one of the DOs.

Thus, using the adaptive/iterative process and the introduced empirical criteria, the wavefront aberrations are compensated. The criterion for the iterative process completion is the formation of a pronounced correlation peak in one of the DOs of the focal plane of the filter, i.e., when the MSE defined by Equation (17) equals a minimal value.

4. Discussion

In order to show the possibility of using the proposed method for detecting wavefronts with complex aberrations, we investigate how much one wave aberration (n_1, m_1) with a level α_1 affects the wave aberration (n_2, m_2) with a level α_2. For an objective assessment, the analysis of the PSF corresponding to the superposition of aberrations is proposed.

Figure 12 shows the simulation results for the superposition of even- and odd-type aberrations. It is seen that if the ratio of the level of aberrations is more than 10 times, then the effect of an aberration with a lower weight coefficient on the resulting PSF is extremely weak. In addition, the ratio of the level of aberrations being more than 5 times also insignificantly affects the resulting wavefront (with an accuracy to the scale and overall shape of the PSF).

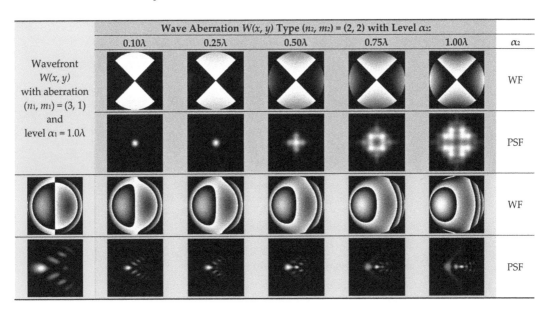

Figure 12. The results of modeling the superposition of the odd wave aberration of the type $(n_1, m_1) = (3, 1)$ with $\alpha_1 = 1.0\lambda$ level (green highlights) and even wave aberrations of the type $(n_2, m_2) = (2, 2)$ with different levels of α_2 (orange highlights). Results of compensation (blue highlights) are placed at the intersection of corresponding rows and columns.

Thus, if the superposition contains aberrations with coefficients different in magnitude, then it makes sense to first compensate for aberration with a large level. This fact allows us to say that the analyzed wavefront with a pronounced single aberration can be successfully detected by the proposed filters defined by Equations (13)–(15) without using an additional iterative algorithm.

5. Conclusions

In this work, we experimentally tested a diffractive multichannel wavefront sensor matched to phase distributions in the form of Zernike functions.

Several variants of filters were designed: filters matched with different types of aberrations of the same level, filters matched with one type of aberration of different levels, and a hybrid filter. Each of the options is convenient for a specific task. It makes sense to apply a hybrid filter for preliminary analysis and then apply a filter matched to different levels for adjusting. When using a dynamic transparency, it is possible to quickly adjust the filters and compensate for the aberration step by step.

Criteria of the proposed adaptive algorithm for determining the aberrations of the analyzed wavefront were formulated: scattering region, PSF compactness, maximum intensity, and orientation of each diffraction order in the focal plane of the multichannel filter. An adaptive compensation of wavefront aberrations was performed using the proposed method and introduced empirical criteria. We show the possibility to reduce a problem dimension at each stage, so the wavefront aberration compensation is provided. We realized an experimental investigation of the proposed approach and applied it for a step-by-step compensation of a complex type of aberrations.

Finally, based on numerical modeling and the results of an optical experiment using a dynamic transparency, we show the efficiency of the proposed filters for detecting wavefront aberrations in a wide range (from 0.1λ to λ).

Considering the wide range of correctly detected aberration values using the proposed method, the field of application can include the measurement and correction of wavefront aberrations in devices such as ground-based telescopes and optical microscopes, in optical communication systems, in industrial laser technology, and in ophthalmology.

Author Contributions: Conceptualization: S.N.K. and P.A.K.; methodology, S.N.K. and P.A.K.; software, P.A.K.; validation, S.N.K., P.A.K. and A.P.P.; formal analysis, S.N.K. and P.A.K.; investigation, S.N.K., P.A.K. and A.P.P.; resources, S.N.K.; data curation, S.N.K., P.A.K. and A.P.P.; writing—original draft preparation, S.N.K. and P.A.K.; writing—review and editing, S.N.K. and P.A.K.; visualization, S.N.K., P.A.K. and A.P.P.; supervision, S.N.K.; project administration, S.N.K.; funding acquisition, S.N.K. All authors have read and agreed to the published version of the manuscript.

Funding: This work was financially supported by the Russian Foundation for Basic Research (grant No. 20-37-90129) for numerical calculations and analysis and by the Ministry of Science and Higher Education of the Russian Federation under the FSRC "Crystallography and Photonics" of the Russian Academy of Sciences (the state task No. 007-GZ/Ch3363/26) for experimental research.

Institutional Review Board Statement: Not applicable.

Informed Consent Statement: Not applicable.

Data Availability Statement: Not applicable.

Conflicts of Interest: The authors declare no conflict of interest.

References

1. Camacho, L.; Mico, V.; Zalevsky, Z.; Garcia, J. Quantitative phase microscopy using defocusing by means of a spatial light modulator. *Opt. Express* **2010**, *18*, 6755–6766. [CrossRef] [PubMed]
2. Lombardo, M.; Lombardo, G. Wave aberration of human eyes and new descriptors of image optical quality and visual performance. *J. Cataract Refract. Surg.* **2010**, *36*, 313–320. [CrossRef] [PubMed]
3. Zhao, Q.; Fan, H.; Hu, S.; Zhong, M.; Baida, L. Effect of optical aberration of telescopes to the laser radar. *Proc. SPIE* **2010**, *7656*, 76565Z.
4. González-Núñez, H.; Prieto-Blanco, X.; De la Fuente, R. Pupil aberrations in Offner spectrometers. *J. Opt. Soc. Am. A* **2011**, *29*, 442–449. [CrossRef] [PubMed]
5. Khonina, S.N.; Ustinov, A.V.; Pelevina, E.A. Analysis of wave aberration influence on reducing focal spot size in a high-aperture focusing system. *J. Opt.* **2011**, *13*, 095702. [CrossRef]
6. Booth, M.; Andrade, D.; Burke, D.; Patton, B.; Zurauskas, M. Aberrations and adaptive optics in super-resolution microscopy. *Microscopy* **2015**, *64*, 251–261. [CrossRef]

7. Khorin, P.A.; Khonina, S.N.; Karsakov, A.V.; Branchevskiy, S.L. Analysis of corneal aberration of the human eye. *Comput. Opt.* **2016**, *40*, 810–817. [CrossRef]
8. Wilby, M.J.; Keller, C.U.; Haert, S.; Korkiakoski, V.; Snik, F.; Pietrow, A.G.M. Designing and testing the coronagraphic modal wavefront sensor: A fast non-common path error sensor for high-contrast imaging. *Proc. SPIE* **2016**, *9909*, 990921.
9. Klebanov, I.M.; Karsakov, A.V.; Khonina, S.N.; Davydov, A.N.; Polyakov, K.A. Wavefront aberration compensation of space telescopes with telescope temperature field adjustment. *Comput. Opt.* **2017**, *41*, 30–36. [CrossRef]
10. Rastorguev, A.A.; Kharitonov, S.I.; Kazanskiy, N.L. Modeling of arrangement tolerances for the optical elements in a spaceborne Offner imaging hyperspectrometer. *Comput. Opt.* **2018**, *42*, 424–431. [CrossRef]
11. Martins, A.C.; Vohnsen, B. Measuring ocular aberrations sequentially using a digital micromirror device. *Micromachines* **2019**, *10*, 117. [CrossRef] [PubMed]
12. Baum, O.I.; Omel'chenko, A.I.; Kasianenko, E.M.; Skidanov, R.V.; Kazanskiy, N.L.; Sobol', E.N.; Bolshunov, A.V.; Avetisov, S.E.; Panchenko, V.Y. Control of laser-beam spatial distribution for correcting the shape and refraction of eye cornea. *Quantum Electron.* **2020**, *50*, 87–93. [CrossRef]
13. Mu, Q.; Cao, Z.; Hu, L.; Li, D.; Xuan, L. Adaptive optics imaging system based on a high-resolution liquid crystal on silicon device. *Opt. Express* **2006**, *14*, 8013–8018. [CrossRef] [PubMed]
14. Ellerbroek, B.L.; Vogel, C.R. Inverse problems in astronomical adaptive optics. *Inverse Probl.* **2009**, *25*, 063001. [CrossRef]
15. Esposito, S.; Riccardi, A.; Pinna, E.; Puglisi, A.; Quirós-Pacheco, F.; Arcidiacono, C.; Xompero, M.; Briguglio, R.; Agapito, G.; Busoni, L.; et al. Large binocular telescope adaptive optics system: New achievements and perspectives in adaptive optics. *Proc. SPIE* **2011**, *8149*, 814902.
16. Lukin, V.P. Adaptive optics in the formation of optical beams and images. *Phys. Uspekhi* **2014**, *57*, 556–592. [CrossRef]
17. Ji, N. Adaptive optical fluorescence microscopy. *Nat. Methods* **2017**, *14*, 374–380. [CrossRef] [PubMed]
18. Bond, C.Z.; Wizinowich, P.; Chun, M.; Mawet, D.; Lilley, S.; Cetre, S.; Jovanovic, N.; Delorme, J.; Wetherell, E.; Jacobson, S.M.; et al. Adaptive optics with an infrared pyramid wavefront sensor. *Proc. SPIE* **2018**, *10703*, 107031Z. [CrossRef]
19. Mahajan, V.N. Zernike circle polynomials and optical aberration of system with circular pupils. *Appl. Opt.* **1994**, *33*, 8121–8124. [CrossRef] [PubMed]
20. Love, G.D. Wavefront correction and production of Zernike modes with a Liquid crystal spatial light modulator. *Appl. Opt.* **1997**, *36*, 1517–1525. [CrossRef] [PubMed]
21. Khonina, S.N.; Kotlyar, V.V.; Soifer, V.A.; Wang, Y.; Zhao, D. Decomposition of a coherent light field using a phase Zernike filter. *Proc. SPIE* **1998**, *3573*, 550–553.
22. Neil, M.A.A.; Booth, M.J.; Wilson, T. New modal wave-front sensor: A theoretical analysis. *J. Opt. Soc. Am. A* **2000**, *17*, 1098–1107. [CrossRef]
23. Booth, M.J. Direct measurement of Zernike aberration modes with a modal wavefront sensor. *Proc. SPIE* **2003**, *5162*, 79–90.
24. Sheppard, C.J.R. Zernike expansion of pupil filters: Optimization of the signal concentration factor. *J. Opt. Soc. Am. A* **2015**, *32*, 928–933. [CrossRef] [PubMed]
25. Porfirev, A.P.; Khonina, S.N. Experimental investigation of multi-order diffractive optical elements matched with two types of Zernike functions. *Proc. SPIE* **2016**, *9807*, 98070E.
26. Wilby, M.J.; Keller, C.U.; Snik, F.; Korkiakoski, V.; Pietrow, A.G.M. The coronagraphic Modal Wavefront Sensor: A hybrid focal-plane sensor for the high-contrast imaging of circumstellar environments. *Astron. Astrophys.* **2017**, *597*, A112. [CrossRef]
27. Khonina, S.N.; Karpeev, S.V.; Porfirev, A.P. Wavefront aberration sensor based on a multichannel diffractive optical element. *Sensors* **2020**, *20*, 3850. [CrossRef]
28. Gerchberg, R.; Saxton, W. Phase determination for image and diffraction plane pictures in the electron microscope. *Optik* **1971**, *34*, 275–284.
29. Fienup, J.R. Reconstruction of an object from the modulus of its Fourier transform. *Opt. Lett.* **1978**, *3*, 27–29. [CrossRef]
30. Elser, V. Phase retrieval by iterated projections. *J. Opt. Soc. Am. A* **2003**, *20*, 40–55. [CrossRef] [PubMed]
31. Marchesini, S.A. Unified evaluation of iterative projection algorithms for phase retrieval. *Rev. Sci. Instrum.* **2007**, *78*, 011301. [CrossRef] [PubMed]
32. Cheng, Z.; Meiqin, W.; Qianwen, C.; Dong, W.; Sui, W. Two-step phase retrieval algorithm using single-intensity measurement. *Int. J. Opt.* **2018**, *2018*, 8643819. [CrossRef]
33. Tokovinin, A.; Heathcote, S. DONUT: Measuring optical aberrations from a single extrafocal image. *Publ. Astron. Soc. Pac.* **2006**, *118*, 1165–1175. [CrossRef]
34. Guo, H.; Korablinova, N.; Ren, Q.; Bille, J. Wavefront reconstruction with artificial neural networks. *Opt. Express* **2006**, *14*, 6456–6462. [CrossRef] [PubMed]
35. Paine, S.W.; Fienup, J.R. Machine learning for improved image-based wavefront sensing. *Opt. Lett.* **2018**, *43*, 1235–1238. [CrossRef] [PubMed]
36. Rivenson, Y.; Zhang, Y.; Günaydın, H.; Teng, D.; Ozcan, A. Phase recovery and holographic image reconstruction using deep learning in neural networks. *Light Sci. Appl.* **2018**, *7*, 17141. [CrossRef]
37. Nishizaki, Y.; Valdivia, M.; Horisaki, R.; Kitaguchi, K.; Saito, M.; Tanida, J.; Vera, E. Deep learning wavefront sensing. *Opt. Express* **2019**, *27*, 240–251. [CrossRef]

38. Rodin, I.A.; Khonina, S.N.; Serafimovich, P.G.; Popov, S.B. Recognition of wavefront aberrations types corresponding to single Zernike functions from the pattern of the point spread function in the focal plane using neural networks. *Comput. Opt.* **2020**, *44*, 923–930. [CrossRef]
39. Khorin, P.A.; Dzyuba, A.P.; Serafimovich, P.G.; Khonina, S.N. Neural networks application to determine the types and magnitude of aberrations from the pattern of the point spread function out of the focal plane. *J. Phys. Conf. Ser.* **2021**, *2086*, 012148. [CrossRef]
40. Khorin, P.A. Analysis wavefront propagating in free space based on the Zernike polynomials and Gauss-Laguerre modes expansion. *J. Phys. Conf. Ser.* **2019**, *1096*, 012104. [CrossRef]
41. Kirilenko, M.S.; Khorin, P.A.; Porfirev, A.P. A wavefront analysis based on Zernike polynomials. *CEUR Workshop Proc.* **2016**, *1638*, 66–75.
42. Khorin, P.A.; Degtyarev, S.A. Wavefront aberration analysis with a multi-order diffractive optical element. *CEUR Workshop Proc.* **2017**, *1900*, 28–33.
43. Khorin, P.A.; Volotovskiy, S.G. Analysis of the threshold sensitivity of a wavefront aberration sensor based on a multi-channel diffraction optical element. *Proc. SPIE* **2021**, *11793*, 117930B.
44. Khorin, P.A.; Podlipnov, V.V.; Khonina, S.N. Generation of scalable wavefront for testing optical systems. *Proc. SPIE* **2020**, *11516*, 115161K.
45. Born, M.; Wolf, E. *Principles of Optics: Electromagnetic Theory of Propagation, Interference and Diffraction of Light*, 7th ed.; Cambridge University Press: Cambridge, UK, 1999.
46. Weaver, C.S.; Goodman, J.W. A technique for optically convolving two functions. *Appl. Opt.* **1966**, *5*, 1248–1249. [CrossRef] [PubMed]
47. Horner, J.L.; Gianino, P.D. Phase-only matched filtering. *Appl. Opt.* **1984**, *23*, 812–816. [CrossRef]
48. Millán, M.S. Advanced optical correlation and digital methods for pattern matching—50th anniversary of Vander Lugt matched filter. *J. Opt.* **2012**, *14*, 103001. [CrossRef]
49. Khonina, S.N.; Karpeev, S.V.; Paranin, V.D. A technique for simultaneous detection of individual vortex states of Laguerre-Gaussian beams transmitted through an aqueous suspension of microparticles. *Opt. Lasers Eng.* **2018**, *105*, 68–74. [CrossRef]
50. Khonina, S.N.; Karpeev, S.V.; Butt, M.A. Spatial-light-modulator-based multichannel data transmission by vortex beams of various orders. *Sensors* **2021**, *21*, 2988. [CrossRef] [PubMed]
51. Khorin, P.A.; Volotovskiy, S.G.; Khonina, S.N. Optical detection of values of separate aberrations using a multi-channel filter matched with phase Zernike functions. *Comput. Opt.* **2021**, *45*, 525–533. [CrossRef]
52. Khorin, P.A. Iterative algorithm for wavefront correction based on optical decomposition in wave aberrations. In Proceedings of the 2021 International Conference on Information Technology and Nanotechnology (ITNT), Samara, Russia, 20–24 September 2021; pp. 1–6.

Article

Focusing of a Laser Beam Passed through a Moderately Scattering Medium Using Phase-Only Spatial Light Modulator

Ilya Galaktionov *, Alexander Nikitin, Julia Sheldakova, Vladimir Toporovsky and Alexis Kudryashov

Institute of Geosphere Dynamics, Leninskiy Avenue 38, Bld. 1, Moscow 119334, Russia; geospheres@idg.chph.ras.ru or nikitin@activeoptics.ru (A.N.); sheldakova@nightn.ru (J.S.); topor@activeoptics.ru (V.T.); kud@activeoptics.ru (A.K.)
* Correspondence: galaktionov@activeoptics.ru

Abstract: The rarely considered case of laser beam propagation and focusaing through the moderately scattering medium was researched. A phase-only spatial light modulator (SLM) with 1920 × 1080 pixel resolution was used to increase the efficiency of focusing of laser radiation propagated through the 5 mm layer of the scattering suspension of 1 μm polystyrene microbeads in distilled water with the concentration values ranging from 10^5 to 10^6 mm^{-3}. A CCD camera with micro-objective was used to estimate the intensity distribution of the far-field focal spot. A Shack-Hartmann sensor was used to measure wavefront distortions. The conducted experimental research demonstrated the 8% increase in integral intensity and 16% decrease in diameter of the far-field focal spot due to the use of the SLM for laser beam focusing.

Keywords: laser beam focusing; scattering medium; spatial light modulator; Shack-Hartmann sensor

1. Introduction

A medium is considered to be turbid or scattering if it has a pronounced optical inhomogeneity due to the presence of impurities of particles with a refractive index that differs from a medium one. Striking examples are atmospheric aerosol, haze, fog, sea water, clouds, biological tissues, etc. [1]. There are at least two reasons why the light beam loses energy while propagating through a turbid medium—absorption and scattering. If the medium absorbs light then the energy is lost, but if the medium scatters light, the energy is not lost, it is just redistributed in space. As a result, it does not allow efficiently focusing the beam on the target and also it makes the contours of objects look blurred. Obviously, this is an obstacle for such applications as pattern recognition, wireless data transmission, medical noninvasive diagnosis, and some others [2–8].

A number of techniques were developed to overcome the problem of imaging and focusing through a scattering medium [5,6]. Clear images of an object could be obtained using holographic techniques by reversing a scattering process [9]. The technique called multispectral multiple-scattering low-coherence interferometry [10] uses coherence and spatial gating to produce images of optical properties of a tissue up to 9 mm deep with millimeter-scale resolution. When there is no access to the space behind a scattering layer, a non-invasive approach to image fluorescent objects can be used [11]. In order to focus light or to reconstruct images of an object placed within or behind a strongly scattering medium, spatial light modulators are often used [7,8,12,13]. In [14], authors demonstrated steady-state focusing of coherent light through dynamic scattering media. In [15], authors demonstrated numerically and experimentally that by using a transmission matrix inversion method peak-to-background intensity ratio of the focus can be higher than that achieved by conventional methods. The authors of [16] developed the new optical time-reversal focusing technique, called time reversal by analysis of changing wavefronts from kinetic targets, that allowed obtaining a focal peak-to-background strength of 204. In [17], authors developed the first full-polarization digital optical phase conjugation system that, on average, doubles the focal peak-to-background ratio achieved by single-polarization digital optical phase conjugation systems. The authors of [18] introduced a phase control technique using programmable acoustic optic deflectors to image the objects in dynamically changing living biological tissue. However, spatial light modulators are widely used in different areas of research; their use is not limited to the imaging and focusing applications only. Spatial light modulators also show vast potential in such applications as micro- and nano-scale fabrication [19].

On the one hand, the radiation wavefront could in principle be correctly measured and optimized with the help of conventional adaptive optics techniques [20–27] in case of the smooth and continuous change of the refractive index of the turbid medium. On the other hand, the radiation wavefront is completely scrambled in case of the scattering medium with random inhomogeneities of high concentration (i.e., biological tissues), and in this special case the wavefront shaping technique could be applied [28].

In other words, there are two regimes that can be distinguished. One regime is when the refractive index of an inhomogeneous medium changes smoothly and continuously in space (optical density value is less than 1). The other regime is when the medium has high concentration of random inhomogeneities (optical density value is higher than 10). The regimes mentioned above can be considered as different ends of a continuum [29], and most of the existing research is dedicated to these particular regimes.

In contrast, in this particular paper we would like to talk about the crossover regime [20]. In this regime we could measure the so-called averaged wavefront (that is not completely scrambled yet) of partially coherent light that has passed through the scattering layer, and we could try to apply the techniques of conventional adaptive optics to improve laser beam focusing (Figure 1).

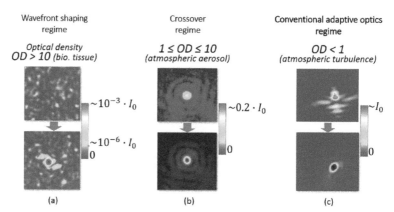

Figure 1. Intensity distributions of the focal spot at the camera sensor placed at the focal plane of a focusing lens for different optical density values of the medium: (**a**) Wavefront shaping regime (image adapted from I. Vellekoop, *Optics Letters* 32(16), 2007); (**b**) crossover regime; (**c**) conventional adaptive optics regime (image adapted with permission from J. Sheldakova et.al., *Proc. SPIE* 4969, 2003).

The uniqueness of the crossover regime is that the focal spot obtained on the imaging sensor plane preserves its spatial structure despite huge energy loss due to scattering. The importance of researching the impact of the scattering effect on the light characteristics in the crossover regime can be explained by the fact that the scattering of the optical inhomogeneities prevents effective wireless energy/information transmission through the moderately scattering media (i.e., atmospheric aerosol, top layers of biological tissue). There are plenty of applications, both imaging and focusing, that will benefit from solving this particular problem—identification of objects in the atmosphere and under the water, increase of visibility of signaling lights of air strips, optical coherence and diffusive tomography, increase of efficiency of free-space optical communication systems, energy transmission to hard-to-reach objects, charging of unnamed aerial vehicles, diagnosis of malignant tissues, monitoring of drug impacts.

The main idea of our work was to control the direction of propagation of scattered (non-ballistic) light in order to improve the efficiency of light focusing [21,22]. Actually, the goal was to decrease the size and increase the energy inside the far-field focal spot obtained in the focal plane of the lens.

To achieve this, we developed a model based on the Shack-Hartmann technique to estimate the distortions of the averaged wavefront of the scattered laser beam and confirmed the simulation results experimentally; we assembled the experimental setup for laser beam focusing through the scattering suspension of polystyrene microspheres, and demonstrated the focusing enhancement.

2. Materials and Methods

2.1. Monte Carlo Simulation of Radiation Transfer through a Scattering Medium

Prior to experimental research on how to minimize the effect of scattering on the quality of a focal spot, we performed a set of numerical simulations. First of all, we simulated the laser beam propagation through a scattering medium that is generally described by the radiative transfer equation [30]. From the wide variety of methods to solve this equation [31], we decided to use the stochastic Monte Carlo technique [32]. For the problem of light propagation, Monte Carlo technique takes into account the quantum nature of light and simulates the behavior of a photon flux [33].

There are three components of a scattered photon flux [34]. *Ballistic* photons travel through a turbid medium without interaction with the scatterers and do not change the initial trajectory. *Quasi-ballistic* photons undergo few scattering events and travel in near-forward paths along a trajectory that is close to the initial direction of the beam propagation. These photons play an important role in imaging and focusing when the thickness of the scattering medium layer increases, because the number of ballistic photons decreases exponentially in this case. *Diffusive* photons scatter in all directions and form a noncoherent component of the scattered light (Figure 2a). Based on the Monte Carlo simulation performed [35,36], we found that there is a big portion of quasi-ballistic photons that passes through the turbid medium and reaches the detector plane (Figure 2b).

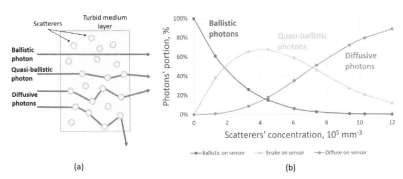

Figure 2. (**a**) Schematic trajectories of photons propagating through a scattering medium: ballistic, quasi-ballistic, and diffusive photons. (**b**) Number of photons reaching the sensor for the scatterers' concentration range from 10^5 to 10^6 mm^{-3} (numerical simulation results).

A big portion of quasi-ballistic photons can be explained by the anisotropic Mie scattering which takes place for the considered scatterers' concentration values (the crossover regime). The idea was to adjust the direction of this quasi-ballistic component by correcting for the overall direction diagram of the scattered beam by means of adaptive optics. In order to do that, it was necessary to find out the influence of the scattering medium on the characteristics of the laser beam. For this, we used the Shack-Hartmann principle [37].

2.2. Shack-Hartmann Principle and Averaged Wavefront Concept

Shack-Hartmann principle consists in dividing the wavefront into a large number of small sub-apertures using a microlens array and measuring the displacements of focal spots in the detector plane. The spot displacement is proportional to the slope of the wavefront at the microlens aperture (Figure 3). Thus, by measuring the spot displacements and calculating the values of the derivatives of the polynomials used (Zernike, B-spline [38], etc.—in this paper we used Zernike), we can calculate the coefficients of the polynomials, which then can be used to approximate the wavefront.

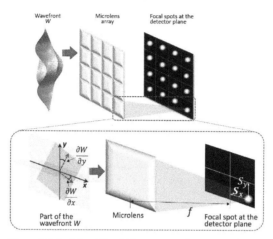

Figure 3. Principle of Shack-Hartmann technique—calculation of wavefront slopes based on focal spots displacements.

Usually, the Shack-Hartmann sensor measures the wavefront of the light, but a laser beam that is passed through the scattering medium in the strict physical sense does not have a wavefront, since part of the radiation is scattered. Initially, before the scattering medium, the beam has a plane wavefront. The scatterers inside the medium that interacts with the radiation become the sources of new spherical waves with their own wave vectors. At the same time, part of the radiation passes through the medium without scattering, keeping the original flat wavefront. As a result, a focal spot is formed at the focal plane of the microlens. The displacement of this focal spot from the optical axis of the microlens is proportional to the local slope of a so-called averaged wavefront. Basically, this averaged wavefront is just a superposition of independent wavefronts from the scatterers and the wavefront of unscattered radiation passed through the medium.

In other words, the averaged wavefront is the superposition of the independent wavefronts coming from each scatterer that acts as a point source (quasi-ballistic and diffusive components) and the wavefront of unscattered (ballistic component) light. This method of measuring the averaged wavefront was implemented in the numerical model [39]. The photons passed through the medium, fell on the aperture of the simulated microlens array and retraced to the receiver plane (geometrical optics approach), forming a set of focal spots called a hartmannogram.

In each sub-aperture (that corresponds to the particular microlens) of the receiver plane, a single focal spot was registered during simulation. The intensity distribution of each focal spot depended on the number of photons registered inside each sub-aperture. The example of the resultant simulated field of focal spots is presented in Figure 4b. The inner blue circle corresponds to the initial diameter of the collimated beam before entering the scattering volume. The outer blue circle corresponds to the area of the Shack-Hartmann sensor where the averaged wavefront was analyzed.

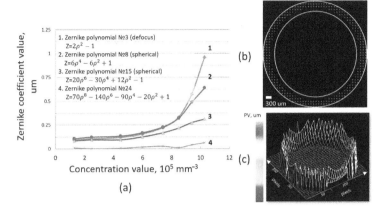

Figure 4. (a) Dependence of Zernike coefficients of symmetric distortions (defocus and spherical aberrations) on the concentration value. (b) Simulated hartmannogram. White dots are focal spots generated by the photons that passed through the scattering medium, fell on the microlens array and were registered on the sensor plane. Inner blue circle shows the initial beam diameter (before it enters the scattering medium), outer blue circle shows the region where new focal spots appear due to the beam scattering. (c) A typical view of the surface of the averaged wavefront. PV ("peak-to-valley") parameter shows an amplitude of averaged wavefront distortions.

To verify the model, an experimental setup was assembled with a diode laser, a suspension of polystyrene microbeads, and Shack-Hartmann sensor (1/2-inch sensor with the size of the receiving area equal to 6.4 × 4.8 mm, focal length of the microlens array— 6 mm, diameter of a single microlens—150 µm, total number of microlenses—greater than 1300). An analysis of the hartmannograms obtained numerically and experimentally showed that the displacements of the focal spots in the central part of the beam were rather small due to the influence of ballistic photons in this area. However, scattering led to the appearance of peripheral focal spots on the sensor outside the diameter of the initial beam, and they were predominantly displaced from the center (Figure 4b). The typical 3D view of the averaged wavefront reconstructed by means of Zernike polynomials is presented in Figure 4c. We obtained such a surface for each of the considered concentration values, and the only difference between these surfaces was an amplitude —the shape of the surface was the same.

A typical view of averaged wavefront surfaces approximated by Zernike polynomials is shown in Figure 4c. The surfaces for different values of the concentration looks very similar and differs only in amplitudes. Since the Mie scattering by spherical particles has a symmetric nature, the distortions were also represented only by centrally symmetric Zernike polynomials. What is important is that in addition to defocus (Zernike polynomial # 3), higher-order spherical distortions were also observed (Figure 4a). Moreover, with an increase in concentration, the amplitude of distortions increased. The tendency towards an increase in the total amplitude of distortions with increasing concentration persisted both in the model and in the experiment [39].

It can be clearly seen that with an increase in the concentration of scatterers, the energy redistribution in the beam as well as the amplitude of the averaged wavefront also increases. As is known, one of the possible ways to control the intensity distribution is to use adaptive optical devices—either deformable mirrors or spatial light modulators [40–42]. The results of the successful use of deformable mirrors based on a bimorph piezoelectric element to improve focusing of a laser beam passed through the moderately scattering medium are presented in [39]. In this paper, we would like to show the results obtained in the similar experimental setup but with the use of a phase-only spatial light modulator.

2.3. Experimental Scheme

The experimental setup for measuring the averaged wavefront distortions and focusing a scattered laser beam was assembled. The scheme of this setup is presented in Figure 5.

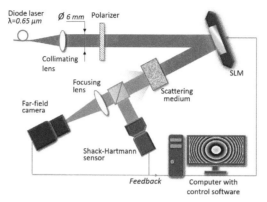

Figure 5. Scheme of the experimental setup with the SLM for laser beam focusing through a scattering medium.

A collimated laser beam (wavelength 0.65 µm) of 6 mm diameter propagated through the polarizer, reflected from the SLM and fell on the transparent 5 mm-thick glass cuvette filled with the polystyrene microbeads of 1 µm diameter diluted in distilled water. The scattered beam fell on the beam-splitter. Part of the beam was directed to the Shack-Hartmann sensor to measure the averaged wavefront. Another part of the beam was focused on the CCD camera with the micro-objective to analyze the intensity distribution of the focal spot. The camera provided image data to the computer. We used the reflective SLM made by Jasper Display Corp. (1920 × 1080 pixels, 12.5 × 7.1 mm active area). It operated in an 8-bit regime. The Shack-Hartmann sensor was based on the CMOS $\frac{1}{2}''$ camera and in this setup was used for measurement purposes only. The CCD camera with $\frac{1}{2}''$ sensor was used as an intensity analyzer.

The flow map of the algorithm that we used for laser beam focusing by means of SLM is presented in Figure 6.

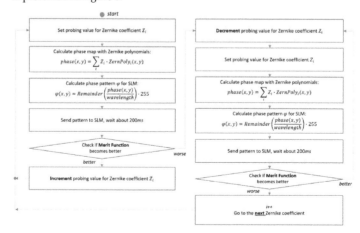

Figure 6. Flow map of the algorithm used to focus a scattered laser beam with the spatial light modulator.

As was shown in previous research [39], low- and high-order axial-symmetric distortions predominated in the averaged wavefront of scattered light. Thus, we decided to operate the SLM using a "phase screen" approach. Instead of controlling each pixel of the SLM individually, we calculated the whole phase surface $phase(x,y)$ of the particular Zernike polynomial and then calculated the phase pattern $\varphi(x,y)$ using the Formula (1) to be sent to the SLM:

$$\varphi(x,y) = 255 \cdot remainder(phase(x,y)/\lambda), \tag{1}$$

where remainder is a reminder of division, $phase(x,y)$ is the phase value at a particular point (X,Y) and λ is the wavelength. The calculated phase pattern was then sent to the SLM. The merit function was calculated after the SLM was settled. In cases where the merit function became better, the current Zernike coefficient was increased by the fixed step (0.015 μm). The whole calculation cycle was run again. In cases where the merit function became worse, the Zernike coefficient was rolled back to the previous value and then decreased. Then the calculation cycle was run again. When the merit function became worse, the best value of the current Zernike coefficient was saved and the algorithm was moved to the next Zernike coefficient and the procedure was repeated. In such a way we improved the merit function step by step. We used the hill-climbing optimization technique; the total number of Zernike polynomials was 36 in this experiment. We used the combination of diameters (Dx, Dy) and peak intensity ($Imax$) of the far-field focal spot as a merit function $M = max(Dx, Dy) \cdot (Dx + Dy)/Imax$.

3. Results and Discussion

Before conducting experiments with the scattering medium, we calibrated the setup. We placed the glass cuvette with the distilled water only (without the polystyrene microbeads) in the optical path and ran the optimization procedure in order to compensate for the induced distortions. We considered the far-field focal spot obtained after optimization as the best achievable focal spot for this experimental setup. In such a way, we took into account all of the misalignments and imperfections of the optical elements of the set-up. After that we measured only the distortions of the laser beam induced by the scattering particles.

After the calibration was done, we started to inject the drops of the scattering dispersion into the cuvette one by one. For each considered concentration of the scattering solution—3.1×10^5 mm^{-3}, 5.1×10^5 mm^{-3}, 6.8×10^5 mm^{-3}, and 8.2×10^5 mm^{-3}—we obtained intensity distributions of the far-field focal spot and corresponding SLM patterns before and after optimization. We also calculated the encircled energy. Figure 7 presents the results of measurements for the minimal (3.1×10^5 mm^{-3}) and maximal (8.2×10^5 mm^{-3}) concentration values considered in this work.

The integral intensity of the focal spot obtained on the imaging sensor when no scattering medium was introduced in the setup was considered as 100%. Then the integral intensities of the focal spots before and after optimization procedure were as follows: 95.7% (before) and 97.7% (after) for the concentration value 3.1×10^5 mm^{-3}, 93.8% and 96.6% for the concentration value 5.1×10^5 mm^{-3}, 85.1% and 91.9% for the concentration value 6.8×10^5 mm^{-3}, 74.3% and 82.2% for the concentration value 8.2×10^5 mm^{-3}. The increase in optimization E was equal to 2.1%, 3%, 8%, and 11%, and was calculated as follows:

$$E = (I_{after} - I_{before})/I_{before} \times 100\%,$$

where I_{before}—integral intensity before optimization, I_{after}—integral intensity of the focal spot after optimization.

The diameters of the focal spots were reduced by 5.2%, 9.7%, 16.4%, and 13.2% during the optimization process, correspondingly.

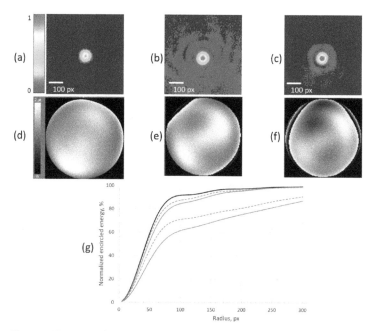

Figure 7. (**a**) Intensity distribution of the far-field focal spot obtained after the calibration procedure was done (glass cuvette contained water only). (**b**,**c**) Intensity distributions after optimization by means of the SLM for the concentration values 3.1×10^5 mm^{-3} and 8.2×10^5 mm^{-3}, correspondingly. (**d**–**f**) SLM patterns that were used to obtain intensity distributions of focal spots presented above. (**g**) Normalized encircled energy charts: black solid curve—without scattering medium; blue curves—with scattering medium with concentration value 3.1×10^5 mm^{-3} before (solid curve) and after (dashed curve) focusing with the SLM; orange curves—with scattering medium with concentration value 8.2×10^5 mm^{-3} before (solid curve) and after (dashed curve) focusing with the SLM.

We note that the decrease of the focal spot diameter for the concentration value of 6.8×10^5 mm^{-3} is lower than the one for the concentration value of 8.2×10^5 mm^{-3} (16.4% versus 13.2%). This is due to the impact of multiply scattered diffuse light. Figure 2b clearly shows that the number of diffuse photons becomes higher than the number of quasi-ballistic and ballistic photons since the concentration value becomes higher than 8×10^5 mm^{-3}. In other words, the integral intensity in the focal spot can become higher due to the optimization, but the diameter of the focal spot is decreasing more slowly because a diffuse light unavoidably broadens the focal spot.

The efficiency of focusing is much higher for the higher concentration value. This is due to the fact that the higher the concentration of scattering medium, the higher the portion of scattered light that can be redirected to the central part of the focal spot using the SLM.

Note that none of the dashed curves in Figure 7 (when the SLM was on) could approach the black solid curve: this is due to energy loss during the scattering process. Part of the scattered laser beam simply did not reach the objective of the imaging camera.

4. Conclusions

In this paper we have demonstrated the use of a phase-only spatial light modulator (SLM) with 1920×1080 pixels resolution for increasing the efficiency of focusing of laser radiation propagated through the moderately scattering medium—5 mm layer of 1 μm polystyrene microbeads diluted in distilled water with concentration values ranging from 10^5 to 10^6 mm^{-3}. We have presented the concept of the averaged wavefront that is the superposition of the independent wavefronts coming from each scatterer that acts as a point source (quasi-ballistic and diffusive components) and the wavefront of unscattered (ballistic component) light. We have demonstrated numerically and experimentally the efficiency of a Shack-Hartmann sensor to measure the averaged wavefront. Experimental research also demonstrated that the SLM allows an increase in the efficiency of laser beam focusing up to 2.1%, 3%, 8%, and 11%, and a decrease in the diameter of the focal spot by 5.2%, 9.7%, 16.4%, and 13.2% for the scatterers' concentration range 10^5 to 10^6 mm^{-3}. Although the increase of focusing efficiency is only up to 11%, it should be noted that we used the phase screen approach during the optimization procedure. Basically, we calculated the phase surface that corresponded to the particular Zernike polynomial and then set it to the SLM rather than controlling each pixel of the SLM individually and one by one (this single pixel approach is rather popular and vastly described in the literature). The advantage of the phase screen approach is a significant increase in optimization speed, the disadvantage is of course lower efficiency. Thus, depending on the problem to be solved one can choose one approach or another. Moreover, promising results can be obtained by combining these approaches.

Author Contributions: Conceptualization, I.G. and A.K.; methodology, I.G., A.N. and A.K.; software, I.G.; validation, A.N., J.S. and I.G.; formal analysis, A.N.; investigation, I.G.; resources, J.S. and V.T.; data curation, A.K.; writing—original draft preparation, I.G.; writing—review and editing, J.S., A.K., V.T. and I.G.; visualization, I.G.; supervision, A.K.; project administration, A.K.; funding acquisition, A.K. All authors have read and agreed to the published version of the manuscript.

Funding: The research was carried out (1) within the state assignment of Ministry of Science and Higher Education of the Russian Federation No. 1021052706254-7-1.5.4—research of laser beam propagation through a scattering medium; (2) within the Russian Science Foundation project # 20-69-46064—research of adaptive optical system control by means of wavefront sensor and intensity analyzer.

Data Availability Statement: The data presented in this study are partially available in ref. [39].

Conflicts of Interest: The authors declare no conflict of interest.

References

1. van de Hulst, H.C. *Light Scattering by Small Particles*; John Wiley & Sons, Inc.: New York, NY, USA, 1957.
2. Mosk, A.P.; Lagendijk, A.; Lerosey, G.; Fink, M. Controlling waves in space and time for imaging and focusing in complex media. *Nat. Photonics* **2012**, *6*, 283. [CrossRef]
3. Bashkatov, A.N.; Priezzhev, A.V.; Tuchin, V.V. Laser technologies in biophotonics. *Quantum Electron.* **2012**, *42*, 379. [CrossRef]
4. Mastiani, B.; Vellekoop, I.M. Noise-tolerant wavefront shaping in a Hadamard basis. *Optics Express.* **2021**, *29*, 2309. [CrossRef] [PubMed]
5. Zhang, Y.; Chen, Y.; Yu, Y.; Xue, X.; Tuchin, V.V.; Zhu, D. Visible and near-infrared spectroscopy for distinguishing malignant tumor tissue from benign tumor and normal breast tissues in vitro. *J. Biomed. Opt.* **2013**, *18*, 077003. [CrossRef] [PubMed]
6. Goodman, J.W.; Huntley, W.H., Jr.; Jackson, D.W.; Lehmann, M. Wavefront reconstruction imaging through random media. *Appl. Phys. Lett.* **1966**, *8*, 311–313. [CrossRef]
7. Katz, O.; Small, E.; Silberberg, Y. Looking around corners and through thin turbid layers in real time with scattered incoherent light. *Nat. Photon.* **2012**, *6*, 549–553. [CrossRef]
8. Popoff, S.M.; Lerosey, G.; Carminati, R.; Fink, M.; Boccara, A.C.; Gigan, S. Measuring the transmission matrix in optics: An approach to the study and control of light propagation in disordered media. *Phys. Rev. Lett.* **2010**, *104*, 100601. [CrossRef]
9. Kogelnik, H.; Pennington, K.S. Holographic imaging through a random medium. *J. Opt. Soc. Am.* **1968**, *58*, 273–274. [CrossRef]
10. Matthews, T.; Medina, M.; Maher, J.; Levinson, H.; Brown, W.; Wax, A. Deep tissue imaging using spectroscopic analysis of multiply scattered light. *Optica* **2014**, *1*, 105. [CrossRef]
11. Bertolotti, J.; van Putten, E.G.; Blum, C.; Lagendijk, A.; Vos, W.L.; Mosk, A.P. Non-invasive imaging through opaque scattering layers. *Nature* **2012**, *491*, 232–234. [CrossRef]
12. Conkey, D.B.; Caravaca-Aguirre, A.M.; Piestun, R. High-speed scattering medium characterization with application to focusing light through turbid media. *Opt. Express* **2012**, *20*, 1733–1740. [CrossRef] [PubMed]
13. Hsieh, C.; Pu, Y.; Grange, R.; Laporte, G.; Psaltis, D. Imaging through turbid layers by scanning the phase conjugated second harmonic radiation from a nanoparticle. *Opt. Express* **2010**, *18*, 20723–20731. [CrossRef] [PubMed]
14. Stockbridge, C.; Lu, Y.; Moore, J.; Hoffman, S.; Paxman, R.; Toussaint, K.; Bifano, T. Focusing through dynamic scattering media. *Opt. Express* **2012**, *20*, 15086–15092. [CrossRef] [PubMed]

15. Xu, J.; Ruan, H.; Liu, Y.; Zhou, H.; Yang, C. Focusing light through scattering media by transmission matrix inversion. *Opt. Express* **2017**, *25*, 27234–27246. [CrossRef]
16. Zhou, E.H.; Ruan, H.; Yang, C.; Judkewitz, B. Focusing on moving targets through scattering samples. *Optica* **2014**, *1*, 227–232. [CrossRef]
17. Shen, Y.; Liu, Y.; Ma, C.; Wang, L.V. Focusing light through scattering media by full-polarization digital optical phase conjugation. *Opt. Lett.* **2016**, *41*, 1130–1133. [CrossRef]
18. Feldkhun, D.; Tzang, O.; Wagner, K.H.; Piestun, R. Focusing and scanning through scattering media in microseconds. *Optica* **2019**, *6*, 72–75. [CrossRef]
19. Li, R.; Jin, D.; Pan, D.; Ji, S.; Xin, C.; Liu, G.; Fan, S.; Wu, H.; Li, J.; Hu, Y.; et al. Stimuli-Responsive Actuator Fabricated by Dynamic Asymmetric Femtosecond Bessel Beam for In Situ Particle and Cell Manipulation. *ACS Nano* **2020**, *14*, 5233–5242. [CrossRef]
20. Barchers, J.D.; Fried, D.L. Optimal control of laser beams for propagation through a turbulent medium. *JOSA A* **2002**, *19*, 1779–1793. [CrossRef]
21. Galaktionov, I.; Kudryashov, A.; Sheldakova, J.; Nikitin, A. Laser beam focusing through the dense multiple scattering suspension using bimorph mirror. *Proc. SPIE* **2019**, *10886*, 1088619.
22. Galaktionov, I.; Kudryashov, A.; Sheldakova, J.; Nikitin, A.; Samarkin, V. Laser beam focusing through the atmosphere aerosol. *Proc. SPIE* **2017**, *10410*, 104100M.
23. Chamot, S.R.; Dainty, C.; Esposito, S. Adaptive optics for ophthalmic applications using a pyramid wavefront sensor. *Optics Express* **2006**, *14*, 518–526. [CrossRef] [PubMed]
24. Roddier, F.J.; Anuskiewicz, J.; Graves, J.; Northcott, M.J.; Roddier, C.A. Adaptive optics at the University of Hawaii I: Current performance at the telescope. *Proc. SPIE* **1994**, *2201*, 2–9.
25. Konyaev, P.A.; Lukin, V.P.; Gorbunov, I.A.; Kulagin, O.V. Hybrid adaptive optical system correcting turbulent distortions on the long atmospheric paths. *Proc. SPIE* **2021**, *11860*, 16–26.
26. Sheldakova, J.; Rukosuev, A.; Alexandrov, A.; Kudryashov, A. Multy-dither adaptive optical system for laser beam control. *Proc. SPIE* **2003**, *4969*, 115–121.
27. Kudryashov, A.; Rukosuev, A.; Samarkin, V.; Galaktionov, I. Fast adaptive optical system for 1.5 km horizontal beam propagation. *Proc. SPIE* **2018**, *10772*, 107720V.
28. Vellekoop, I.M.; Mosk, A.P. Focusing coherent light through opaque strongly scattering media. *Opt. Lett.* **2007**, *32*, 2309–2311. [CrossRef]
29. Vellekoop, I.M. Feedback-based wavefront shaping. *Opt. Express* **2015**, *23*, 12189–12206. [CrossRef]
30. Berrocal, E.; Sedarsky, D.; Paciaroni, M.; Meglinski, I.; Linne, M. Laser light scattering in turbid media Part I: Experimental and simulated results for the spatial intensity distribution. *Opt. Express* **2007**, *15*, 10649–10665. [CrossRef]
31. Vorob'eva, E.A.; Gurov, I.P. *Models of Propagation and Scattering of Optical Radiation in Randomly Inhomogeneous Media*; Moscow Meditsina: Moscow, Russia, 2006.
32. Wang, L.; Jacques, S. MCML—Monte Carlo modeling of light transport in multi-layered tissues. *Comput. Programs Methods Biomed.* **1995**, *47*, 131. [CrossRef]
33. Meglinski, I.V. Quantitative assessment of skin layers absorption and skin reflectance spectra simulation in the visible and near-infrared spectral regions. *Physiol. Meas.* **2002**, *23*, 741. [CrossRef] [PubMed]
34. Ramachandran, H. Imaging through turbid media. *Curr. Sci.* **1999**, *76*, 1334.
35. Galaktionov, I.V.; Kudryashov, A.V.; Sheldakova, Y.V.; Byalko, A.A.; Borsoni, G. Measurement and correction of the wavefront of the laser light in a turbid medium. *Quantum Electron.* **2017**, *47*, 32–37. [CrossRef]
36. Galaktionov, I.; Kudryashov, A.; Sheldakova, J.; Nikitin, A.; Samarkin, V. Laser beam focusing through the scattering medium by means of adaptive optics. *Proc. SPIE* **2017**, *10073*, 100731L.
37. Nikitin, A.; Galaktionov, I.; Denisov, D.; Karasik, V.; Sakharov, A.; Baryshnikov, N.; Sheldakova, J.; Kudryashov, A. Absolute calibration of a Shack-Hartmann wavefront sensor for measurements of wavefronts. *Proc. SPIE* **2019**, *10925*, 109250K.
38. Galaktionov, I.; Nikitin, A.; Sheldakova, J.; Kudryashov, A. B-spline approximation of a wavefront measured by Shack-Hartmann sensor. *Proc. SPIE* **2021**, *11818*, 118180N.
39. Galaktionov, I.; Sheldakova, J.; Nikitin, A.; Samarkin, V.; Parfenov, V.; Kudryashov, A. Laser beam focusing through a moderately scattering medium using bimorph mirror. *Opt. Express* **2020**, *28*, 38061–38075. [CrossRef]
40. Galaktionov, I.; Kudryashov, A.; Sheldakova, J.; Nikitin, A. Laser beam focusing through the scattering medium using bimorph deformable mirror and spatial light modulator. *Proc. SPIE* **2019**, *11135*, 111350B.
41. Sheldakova, J.; Galaktionov, I.; Nikitin, A.; Rukosuev, A.; Kudryashov, A. LC phase modulator vs deformable mirror for laser beam shaping: What is better? *Proc. SPIE* **2018**, *10774*, 107740S.
42. Sheldakova, J.; Toporovsky, V.; Galaktionov, I.; Nikitin, A.; Rukosuev, A.; Samarkin, V.; Kudryashov, A. Flat-top beam formation with miniature bimorph deformable mirror. *Proc. SPIE* **2020**, *11486*, 114860E.

Article

Filtered Influence Function of Deformable Mirror for Wavefront Correction in Laser Systems

Yamin Zheng [1,2], Ming Lei [3], Shibing Lin [1,2], Deen Wang [1,2,4], Qiao Xue [4] and Lei Huang [1,2,*]

[1] Key Laboratory of Photonic Control Technology Ministry of Education, Department of Precision Instrument, Tsinghua University, Beijing 100084, China; yamin.zheng@foxmail.com (Y.Z.); linsb19@mails.tsinghua.edu.cn (S.L.); sduwde@126.com (D.W.)
[2] State Key Laboratory of Precision Measurement Technology and Instruments, Department of Precision Instrument, Tsinghua University, Beijing 100084, China
[3] China Defense Science and Technology Information Center, Beijing 100084, China; minglei_cdstic@163.com
[4] Research Center of Laser Fusion, China Academy of Engineering Physics, P.O. Box 919-988, Mianyang 621900, China; 031205100@163.com
* Correspondence: hl@tsinghua.edu.cn

Abstract: An influence function filtering method (IFFM) is presented to improve the wavefront correction capability in laser systems by curbing the correction performance degradation resulted from the IF measurement noise. The IFFM is applied to the original measured IF. The resulting filtered IF is then used to calculate the wavefront control signal in each iteration of the closed-loop correction. A theoretical wavefront correction analysis model (CAM) is built. The impact of the IF measurement noise as well as the improvement of the IFFM on the wavefront correction capability are analyzed. A simulation is set up to analyze the wavefront correction capability of the filtered IF using Zernike mode aberrations. An experiment is carried out to study the effectiveness of the IFFM under practical conditions. Simulation and experimental results indicate that the IFFM could effectively reduce the negative effect of the measurement noise and improve the wavefront correction capability in laser systems. The IFFM requires no additional hardware and does not affect the correction speed.

Keywords: adaptive optics; influence function; wavefront correction

1. Introduction

Adaptive optics (AO) is widely employed in many fields, from astronomical observation [1–8] at the macro level to bioimaging microscopy [9–15] at the micro level and from inertial confinement fusion facilities [12–14] at the high-power level to vision science research [15–17] at the low-power level. One of the important applications of AO is wavefront correction in laser systems to improve the beam quality of the output laser beam [18–20]. An AO system generally consists of a wavefront sensor, a wavefront controller and a wavefront compensator. The deformable mirror (DM) is commonly adopted as the wavefront compensator [21,22]. In the wavefront correction process, the DM's influence function (IF) and the distorted wavefront to be corrected are measured by the wavefront sensor, and then the measured data are transmitted to the wavefront controller to calculate the control signal of the DM. The surface shape of the DM is then deformed to generate the conjugated wavefront to correct the wavefront aberration. Therefore, the measurement accuracy of the IF should be high enough to ensure the accuracy of the calculated control signal so as to obtain fine correction capability. However, in many practical optical systems, the IF is commonly measured by using a Shack–Hartmann wavefront sensor (SHWFS) and the measurement noises always exists in the measured IF [23–25]. According to previous research, the measurement noises occur in the form of random and irregular fluctuations with amplitude being about 30% of the overall deformation amplitude of the DM caused by ambient vibration and turbulence [23–25]. Due to occurrence of the measurement noises, the measured IF is inconsistent with the actual IF of the DM, and the wavefront correction ability of the AO system is limited.

Some studies have been focused on the improvement of the measurement accuracy of the IF to obtain good wavefront correction capability [26–29]. An adaptive-influence-matrix (AIM) method was proposed, in which the IF was measured and calibrated multiple times during the closed-loop correction process to improve the measurement accuracy [26]. This method took the nonlinearity effect of the actuators into consideration while leaving out the random and irregular measurement noise. Then, a two-step high-precision system identification method in wavefront control was proposed [27] with the measurement noise being considered, in which multiple tentative measured wavefronts were used to suppress the measurement noise and obtain optimized influence functions. Besides, a wavefront reconstruction method was proposed [28] to estimate the high-resolution wavefront using multiple wavefront measurements. The high-resolution wavefront was reconstructed by a maximum a posteriori (MAP) method, taking into consideration the measurement model and the prior information derived from the spectrum statistics of the turbulence phase. The accuracy of the measured IF could hence be improved. Nonetheless, in these methods that could improve the IF measurement accuracy, multiple IF measurements were required, which would reduce the applicability of the AO system. Recently, a hybrid AO system-based IF optimization method was proposed [29], in which the IF was measured by using a deflectometry system with higher measurement accuracy than the SHWFS. Simulation and experimental results revealed that the method could improve the IF measurement accuracy as well as the wavefront correction capability, however the utilization of additional hardware would increase the structure complexity of the system. In addition to this, some studies focused on fitting the actual IF with Gaussian function and its modifications [30,31]. A modified Gaussian influence function (MGIF) was proposed to fit the IF measured by a Zygo interferometer [30], and a simple method is adopted to characterize the actual IF of various position actuators. Compared to the traditional Gaussian IF, the MGIF possesses the actual IF features in the azimuthal and radial directions. Besides, the IF of a 140-actuator continuous membrane DM was measured using an SHWFS and then fitted using a Gaussian function [31]. The Gaussian and modified Gaussian function fitting methods could be implemented in the IF measurement during the precorrection process, which means that the effect of the measurement noise could be mitigated in the correction; therefore, the closed-loop correction capability could be improved. Nonetheless, this requires that the IF needs to be accurately measured in the precorrection process without the measurement noise, thus reducing the flexibility of the measurement.

An IF filtering method (IFFM) to effectively improve the closed-loop wavefront correction capability of the AO system in laser systems is presented in this paper. The IFFM is applied during the calculation process of the wavefront correction. In the IFFM, a filtered IF is generated in the precorrection process based on the measured IF through the filtering function. After that, the filtered IF will replace the measured IF in the closed-loop correction process to calculate the wavefront correction control signal with the measured laser wavefront. The filtered-IF-based control signal is then applied to the DM to improve the closed-loop correction capability of the AO system. This paper is organized as follows. In Section 2, we build a theoretical wavefront correction analysis model (CAM) in simulation to systematically analyze the impact of the measurement noises on the correction capability. We also introduce principles of the IFFM, including the calculation process and the characteristic analysis of the IF before and after the filtering. In Section 3, a simulation model is built to investigate the improvement of correction capability brought by the IFFM. In Section 4, an experiment is conducted to study the effectiveness of the IFFM in practical circumstances. In Section 5, we analyze and discuss the simulation and experimental results, which show that the correction capability as well as the robustness of the AO system could be improved by using the IFFM.

2. Principles
2.1. Principles of the Wavefront Correction Analysis Model

In an AO system, in which a DM is employed as the wavefront compensator and the IF of the DM is taken to calculate the control signals, the measurement noise of the IF has a negative effect on the correction capability of the system. To theoretically investigate the effect of measurement noises, a theoretical CAM is built. This model explains how IF works in the wavefront correction process of an AO system, and clarifies the relationship of the IF, the initial wavefront (which is to be corrected) and the residual wavefront after correction. Symbols used in the CAM and corresponding descriptions are listed in Table 1.

Table 1. Symbols used in the CAM.

Symbol	Description
w	Initial wavefront before correction
r	Residual wavefront after correction
u	Control signal of the DM
F_a	Simulated ideal IF of the DM
F_{measu}	Measured IF of the DM
F_{modu}	Filtered IF of the DM
δ	Measurement noise of the IF

The parameters in the CAM are defined as follows. The initial wavefront w represents the wavefront measured by the wavefront sensor and to be corrected by the AO system. The control signal u represents the voltage signal that is applied on the DM to generate conjugate wavefront to compensate the initial wavefront w. The residual wavefront r represents the uncorrected residual after the correction of the initial wavefront w. The ideal IF F_a represents the IF simulated based on the model. The measured IF F_{measu} represents the DM's IF measured by the wavefront sensor. The filtered IF F_{modu} represents the IF filtered from the measured IF F_{measu} by the IFFM. The IF measurement noise δ represents the difference between the measured IF F_{measu} and the ideal IF F_a.

In the CAM, the wavefront correction process of the AO system is simplified into two steps. In step one, w is set as the compensation target and u is calculated using F_c and w through a calculation algorithm. Here, we adopt the least squares method, a form of mathematical regression analysis often used in AO wavefront correction, to calculate the control signal u that best fits the DM's IF and the wavefront aberration to be corrected [32–34]. In step two, the compensating wavefront is generated using the ideal IF F_a and the control signal u, and the residual r is calculated as the difference between w and the compensating wavefront. The two steps are expressed as Equations (1) and (2):

$$F_c u = -w, \tag{1}$$

$$r = w + F_a u, \tag{2}$$

where F_c represents the IF (i.e., F_a, F_{measu} or F_{modu}) that is used in the calculation of the control signal u. Taking Equation (2) as a linear system of equations, there exists an optimal control signal u_{op} that could minimize the residual, and u_{op} is given by

$$u_{op} = -\left(F_a^T F_a\right)^{-1} F_a^T w. \tag{3}$$

Meanwhile, the solution u_c of u in Equation (1) is given by

$$u_c = -\left(F_c^T F_c\right)^{-1} F_c^T w. \tag{4}$$

It is noteworthy that F_a in Equation (2) represents the ideal IF and could not be accurately measured in experiments since measurement noise δ always exists. Thus, it could be seen that u_{op} and u_c are not identical invariably, which will finally lead to the differences in wavefront correction capability.

F_{measu} is used as F_c in practical correction process:

$$F_c = F_{measu} = F_a + \delta. \qquad (5)$$

If the measurement is ideal and no measurement noise (i.e., $\delta = 0$) exists, F_c and u_c equal the ideal IF F_a and u_{op}, respectively, and the optimal r could be obtained. However, the ideal measurement could not be achieved practically due to ambient vibration and turbulence. The measurement noise δ does not equal zero and F_c equals the measured IF F_{measu}, while u_c is not equal to u_{op}. Thus, the wavefront correction capability is affected.

Previous research shows that for an AO system operating indoors, the IF measurement noise δ appears as random and irregular fluctuations in the whole clear aperture [18–20]. For each actuator's IF, the peak-to-valley (PV) value of the measured noise δ is generally 20–40% of that of the ideal IF F_a. Due to the measurement noise, the optimal r could not be obtained in a practical AO system, which will result in ability degradation of the wavefront correction.

A simulation model is built to investigate the effect of the measurement noise δ on the residual r. In the simulation, the ideal IF F_a is built through the COMSOL Multiphysics software, while the measured IF F_{measu} as well as the correction process are simulated through MATLAB software [35,36].

As shown in Figure 1, a finite element model of the DM, which consists of a base, a mirror and 49 in-between actuators, is built in COMSOL. Structural and material parameters of the DM are listed in Tables 2 and 3 separately. BK7, stainless steel and piezoelectric ceramics are set as materials of the mirror, the base and the actuators, respectively. The model is based on the stress interface in the structural mechanics module. The vertical deformation, horizontal rotation and center point of the base undersurface are all set to zero degrees of freedom. The initial surface shape of the mirror is set in an absolute plane. In the simulation, the top faces of the 49 actuators are set to shift up by 1 μm individually in order to obtain the ideal IF F_a, and the resulting mirror surface shape deformations are recorded as the IFs in forms of matrices accordingly.

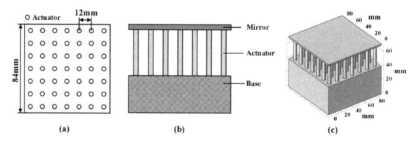

Figure 1. Schematic diagrams of the DM. (**a**) Actuator distribution, (**b**) side view of structure and (**c**) COMSOL model of the DM.

Table 2. Structural parameters of the DM.

Parameter	Mirror	Actuator	Base
Value/mm	84 × 84 × 2	Φ5 × 30	84 × 84 × 40

Table 3. Material parameters of the DM.

Parameter	BK7	Stainless Steel	Piezoelectric Ceramics
Young's Modulus/GPa	81	193	75
Poisson's Ratio	0.17	0.3	0.35
Density/kg·m^{-3}	2400	7930	7750
Linear Expansivity/10^{-6}·K^{-1}	7.1	17.2	12

In the simulation, in order to calculate the measured IF F_{measu}, the IF measurement noise δ is generated and added to the ideal IF F_a in MATLAB software. The measurement noise δ used in the simulation is based on the noise data measured under experimental conditions. The experimental noise data is measured on an optical table equipped with four self-leveling active isolation table legs to avoid strong vibrations, and the fans of the cleanroom are closed to avoid strong air turbulence. The experimentally measured noise data have the characteristics of random distribution as well as low amplitude. Therefore, the noise δ for each actuator is set as continuous and irregular in the whole clear aperture, while the PV value of the noise is set 30% of that of the ideal IF F_a. Thus, for each actuator, the noise $\delta(i)$ is added to the ideal IF $F_a(i)$ (i from 1 to 49) and the measured IF $F_{measu}(i)$ is obtained as

$$F_{measu}(i) = F_a(i) + \delta(i), \quad i = 1, 2, \ldots, 49. \quad (6)$$

For all the 49 actuators, the measurement noises are set random and distinct:

$$\delta(i) \neq \delta(j), \quad i, j = 1, 2, \ldots, 49, \quad i \neq j. \quad (7)$$

Taking the 25th actuator (i.e., the central actuator of the DM) as an example, Figure 2 illustrates its simulated ideal IF $F_a(25)$, measurement noise $\delta(25)$ and measured IF $F_{measu}(25)$. It could be seen that in the simulated ideal IF $F_a(25)$, the surface fluctuation is concentrated in the form of a peak and the height gradually decreases outwards from the center. After the measurement noise $\delta(25)$ is added to $F_a(25)$, the simulated measured IF $F_{measu}(25)$ is obtained with random fluctuations in the whole clear aperture.

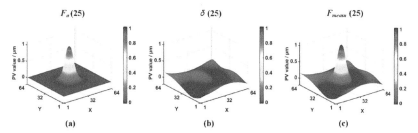

Figure 2. Comparison between F_{measu} and F_a, taking the 25th actuator as example. (a) $F_a(25)$, (b) $\delta(25)$ that is added to the $F_a(25)$, (c) $F_{measu}(25)$.

The Gaussian function has been employed to fit the measured IF in many studies with the expression of

$$G(x, y) = A \cdot exp\left[ln(\omega) \cdot \frac{(x - c_x)^2 + (y - c_y)^2}{2\sigma^2}\right] \quad (8)$$

where A is the peak amplitude, ω is the coupling coefficient, c_x and c_y are the coordinates of the Gaussian function center, and σ is the standard deviation value [30,31]. The generated ideal IF $F_a(25)$ could be fitted by the Gaussian function with the parameters set $A = 0.9841$, $\omega = 0.15$ and $\sigma = 8$. The root-mean-square (RMS) value of the fitting residual is 0.0087 µm, which is 93% smaller than the RMS value of $F_a(25)$ (i.e., 0.1208 µm).

The wavefront correction abilities of the DM using the ideal IF F_a and the measured IF F_{measu} are verified by taking the 3rd to 10th Zernike mode aberrations [37,38] as the initial wavefronts. In step one of the correction process, F_a and F_{measu} are set as F_c (Equation (1)), respectively. In step two, F_a is used to calculate the residual (Equation (2)). PV values and RMS values of the corrected wavefront are depicted in Figure 3a,b, separately. In Figure 3, the blue bars represent the correction results when no measurement noise occurs and F_c equals the ideal IF F_a, while orange bars represent the results when measurement noise exists and F_c equals the measured IF F_{measu}. As shown in Figure 3, the PV values of the corrected wavefront aberrations by using F_a is 31% lower on average than those by using F_{measu}. For the RMS values, the reduction is as large as 47%. This indicates that the measured IF F_{measu} is less effective than the ideal IF F_a in the wavefront correction for the 3rd to 10th Zernike mode aberrations.

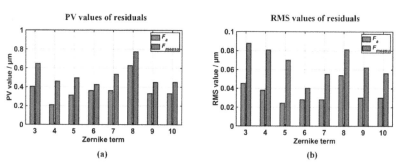

Figure 3. (a) PV and (b) RMS values of correction residuals of the 3rd–10th Zernike mode aberrations. The blue bars represent the correction results by using F_a, and the orange bars by using F_{measu}.

In conclusion, simulation results suggest that F_a has better correction capability than F_{measu}. In other words, the measurement noise δ would result in the degradation of the wavefront correction ability. To reduce the negative effect of the measurement noise δ on the AO system's capability, the IFFM is proposed.

2.2. Principles of IFFM

In the wavefront correction process of an AO system, the DM operates as the wavefront compensator while the measured IF (F_{measu}) of the DM is used along with the initial wavefront w to calculate the control signal u. The IFFM is introduced into step one of the CAM (i.e., the calculation process of the control signal) to restrain the degradation of the wavefront correction capability brought by the measurement noise δ. The IFFM filters the measured IF F_{measu} into the filtered IF F_{modu}, which is used for the calculation of the control signal.

Filtering methods used in the IFFM should abide by two principles: the characteristics of central peak of the measured IF F_{measu} in certain areas around the center should be preserved, and the random noise of the measured IF F_{measu} in the area far from the center should be suppressed. Based on the two principles, three types of filtering methods are proposed, including the cut-type filtering, the linear-type filtering and the Gaussian-type filtering, to reduce the measurement noise δ without losing the characteristics of the central peak. The filtered IF F_{modu} is given by

$$F_{modu}(i) = M(i) \odot F_{measu}(i) \quad i = 1, 2, \ldots, 49. \tag{9}$$

Here, the filtering matrix $M(i)$ [$M_1(i)$, $M_2(i)$ and $M_3(i)$] is used to filter the measured IF $F_{measu}(i)$. The symbol $\odot$ means the element-wise product of matrices, where each element j, k of the new matrix is the product of elements j, k of the original two matrices. The $F_{modu}(i)$ matrix is the element-wise product of the $F_{measu}(i)$ matrix and the $M(i)$ matrix. To compare the filtering matrices, the central actuator of the DM (i.e., the 25th actuator) is taken as an example. Shown in Figure 4 are the element values of $M_1(25)$, $M_2(25)$ and $M_3(25)$. The filtering matrices cover the range of the full DM's surface with a size of 64 by 64 pixels.

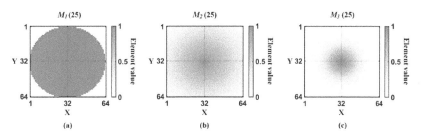

Figure 4. Values of the elements of (**a**) $M_1(25)$ for cut-type filtering, (**b**) $M_2(25)$ for linear-type filtering and (**c**) $M_3(25)$ for Gaussian-type filtering.

The first type is the cut-type filtering. For the filtering matrix $M_1(i)$ of each actuator, a circular area is specified and the elements inside it have the same value of 1, while the elements outside it have the same value of 0. The position of the element with the largest value of $F_{measu}(i)$ is selected as the center of the circle area of $M_1(i)$. Then, the position of the nearest element with the value being reduced by 95% or more of $F_{measu}(i)$ is selected, and the distance between the selected position and the circle's center is taken as the radius of the circular area of $M_1(i)$. It could be seen in Figure 4a that the cut-type filtering could preserve the characteristics in the circular area near the peak of the measured IF F_{measu}, and the characteristics outside the circular area are eliminated.

The second type is the linear-type filtering. For the filtering matrix $M_2(i)$ of each actuator, the elements' values are considered decreasing linearly from 1 to 0 outwards from the specified center. Like $M_1(i)$, the position of the element with the largest value of $F_{measu}(i)$ is selected as the center, and the element of $M_2(i)$ at this position has the value of 1. The element of $M_2(i)$ with the furthest distance from the specified center has the value of 0. The value of each element of $M_2(i)$ depends on the element's distance from the specified center and decreases linearly with the distance. It could be seen in Figure 4b that the linear-type filtering modulates the whole aperture of the measured IF F_{measu}. Compared with other two types, the linear-type filtering preserves all the characteristics of the measured IF $F_{measu}(i)$, including the central peak characteristics (i.e., the ideal IF F_a) and the random fluctuations (i.e., the measurement noise δ).

The third type is the Gaussian-type filtering. For the filtering matrix $M_3(i)$ of each actuator, elements' values are considered decreasing nonlinearly with distance from the specified center and conforming to the Gaussian distribution. Like $M_1(i)$ and $M_2(i)$, the position of the element with the largest value of $F_{measu}(i)$ is selected as the center, and the element of $M_3(i)$ at the same position has the value of 1. The position of the nearest element with the value reduced by 95% or more of $F_{measu}(i)$ is selected, and the distance between the selected position and the specified center is taken as the standard deviation σ of the Gaussian distribution. It could be seen from Figure 4c that the Gaussian-type filtering modulates the whole aperture and partially preserved the fluctuations of the measured IF F_{measu}. However, compared with those in the central area, the fluctuations outside are greatly suppressed.

Filtering results of the measured IF by using these three types of filtering methods are depicted in Figure 5, taking the 25th actuator as an example. Illustrated in Figure 6 are the cutaway views of the surface shape deformations of the IFs.

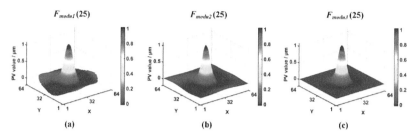

Figure 5. Filtered IFs. (a) Cut-type filtering result $F_{modu1}(25)$. (b) Linear-type filtering result $F_{modu2}(25)$. (c) Gaussian-type filtering result $F_{modu3}(25)$.

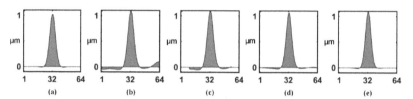

Figure 6. Cutaway views of the surface shape deformation of (a) $F_a(25)$, (b) $F_{measu}(25)$, (c) $F_{modu1}(25)$, (d) $F_{modu2}(25)$, and (e) $F_{modu3}(25)$.

It could be seen that random fluctuations of the IF filtered by the cut-type filtering (F_{modu1}) are eliminated. The IF filtered by the linear-type filtering (F_{modu2}) has a linear peak and the random fluctuations outside the central area are depressed, while the IF filtered by the Gaussian-type filtering (F_{modu3}) has a most concentrated peak and the random fluctuations outside the central area are eliminated. To investigate the difference between F_{modu1}, F_{modu2}, F_{modu3} and F_{measu}, the IF deviation characteristics (Δ) from the ideal IF F_a are given by

$$\begin{cases} \Delta_{measu} = F_{measu} - F_a \\ \Delta_{modu1} = F_{modu1} - F_a \\ \Delta_{modu2} = F_{modu2} - F_a \\ \Delta_{modu3} = F_{modu3} - F_a \end{cases}. \tag{10}$$

Here, Δ is defined as the difference between the IF F_c and the ideal IF F_a. Each IF deviation Δ consists of $\Delta(i)$ for all the 49 actuators. To evaluate the deviation between the IF F_c and the ideal IF F_a, the RMS values of the IF deviation $\Delta(i)$ are calculated for each actuator. Under ideal measurement conditions, the IF F_c equals the ideal IF F_a, which means the IF deviation Δ is 0. Otherwise, the IF F_c is not equal to the ideal IF F_a, and the RMS values of the IF deviation $\Delta(i)$ could be used to evaluate the difference.

As can be seen in Figure 7, for all of the 49 actuators, the RMS values of the IF deviation $\Delta_{measu}(i)$ range from 0.038 to 0.095 µm, while those of the $\Delta_{modu1-3}(i)$ from 0.011 to 0.060 µm. For each actuator, the RMS value of the IF deviation $\Delta_{measu}(i)$ is larger than that of the IF deviation $\Delta_{modu1-3}(i)$. The difference is even larger than 50% for some actuators (e.g., 6th, 16th and 45th actuators). This indicates that the IF deviation between the measured IF F_{measu} and the ideal IF F_a is larger than that between the filtered IFs $F_{modu1-3}$ and the ideal IF F_a. In other words, the deviation between the measured IF and the ideal IF F_a brought by the measurement noise δ could be effectively reduced by using the IFFM.

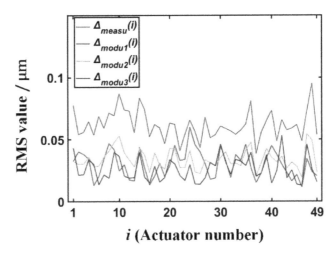

Figure 7. RMS values of the IF deviation $\Delta_{measu}(i)$, $\Delta_{modu1}(i)$, $\Delta_{modu2}(i)$ and $\Delta_{modu3}(i)$ for each actuator.

3. Simulation

A simulation is set up to further investigate the improvement by the IFFM on the wavefront correction capability of an AO system. F_{measu}, F_{modu1}, F_{modu2} and F_{modu3} are set as F_c in step one of the CAM to, respectively, calculate the control signal u (Equation (1)), which is then used along with the ideal IF F_a to accomplish the wavefront compensation and calculate the residual r in step two. The 3rd to 10th Zernike mode aberrations, as well as an actual wavefront measured in the experiment, are set as the initial wavefront w, respectively. The PV and RMS values of the correction residuals of the Zernike mode aberrations are displayed in Figure 8.

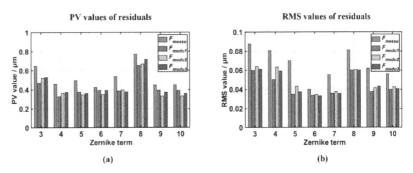

Figure 8. (a) PV values and (b) RMS values of correction residuals of the 3rd–10th Zernike mode aberrations by using F_{measu}, F_{modu1}, F_{modu2}, and F_{modu3}.

As shown in Figure 8, for the 3rd to 10th Zernike mode aberrations, the PV and RMS values of the residual r by using F_{modu} are, on average, 20% and 30% better than those by using F_{measu}, respectively. For instance, for the 6th Zernike mode aberrations, the PV values of the residual r by using $F_{modu1-3}$ (0.393, 0.349 and 0.391 μm) are 11% lower on average than those by F_{measu} (0.425 μm), while the RMS values by using $F_{modu1-3}$ (0.034, 0.035 and 0.033 μm) are 15% lower on average than those by using F_{measu} (0.040 μm). As for the 7th Zernike mode aberrations, the PV values of the residual r by using $F_{modu1-3}$ (0.386, 0.397 and 0.378 μm) are 28% lower on average than those by using F_{measu} (0.538 μm), while the RMS values by using $F_{modu1-3}$ (0.036, 0.038 and 0.036 μm) are 33% lower on average than those by F_{measu} (0.055 μm). It should be noted that the PV and RMS values of the residual r by using F_{modu1}, F_{modu2} and F_{modu3} have little difference. As shown in Figure 8, the maximum difference between the PV values of the residual r by using the three filtered IFs is 16% at the 9th Zernike mode aberration correction, while the maximum difference between the RMS values is 50% at the 5th Zernike mode aberration correction. From Figure 8a it could be seen that F_{modu1} achieves the lowest PV value of correction residuals for the 3rd, 4th and 8th Zernike mode aberrations, F_{modu2} achieves the lowest PV value of correction residuals for the 5th, 6th, 9th and 10th Zernike mode aberrations, while F_{modu3} performs the best for the 7th Zernike mode aberrations. From Figure 8b it could be seen that in terms of the RMS values of correction residuals, F_{modu1} performs the best for the 3rd to 5th and the 8th to 10th Zernike mode aberrations, while F_{modu3} performs the best for the 6th and 7th Zernike mode aberrations. It should be noticed that for the 5th and 7th Zernike mode aberrations, the difference of RMS values of correction residuals by using F_{modu1} and F_{modu3} are less than 3% and 1%, respectively. The result indicates that for the correction of the 3rd to 10th Zernike mode aberrations, $F_{modu1-3}$ perform similarly and better than F_{measu}.

To further investigate the effect of the IFFM on the wavefront correction capability in simulation, in addition to the Zernike mode aberrations, an actual wavefront measured in the experiment is taken as the initial wavefront w (Figure 9a). Figure 9b shows that the initial wavefront mainly contains the 13th, 22nd and 24th Zernike mode aberrations. Illustrated in Figure 9c–g are the correction results by using F_{measu}, F_{modu1}, F_{modu2} and F_{modu3}, respectively. As shown in Figure 9d, the residual r by using F_{measu} has the largest PV value of 0.671 μm and the largest RMS value of 0.094 μm. The PV values of the residual r by using F_{modu1} (0.475 μm), F_{modu2} (0.518 μm) and F_{modu3} (0.482 μm) are 29%, 23% and 28% lower than those by F_{measu}, while the RMS values by using F_{modu1} (0.064 μm), F_{modu2} (0.078 μm) and F_{modu3} (0.070 μm) are 32%, 18% and 25% lower than those by F_{measu}. This difference indicates that the correction results by using F_{modu1}, F_{modu2} and F_{modu3} are better than those by using F_{measu}. Furthermore, the surface shape of the residual by using $F_{modu1-3}$ (Figure 9e–g) are smoother than that by using F_{measu} (Figure 9c). Among the three filtered IFs, F_{modu1} achieves the lowest PV and RMS values of correction residuals. $F_{modu2-3}$ perform worse but still better than F_{measu}. In conclusion, the wavefront correction capability of the AO system could be effectively improved by using the IFFM.

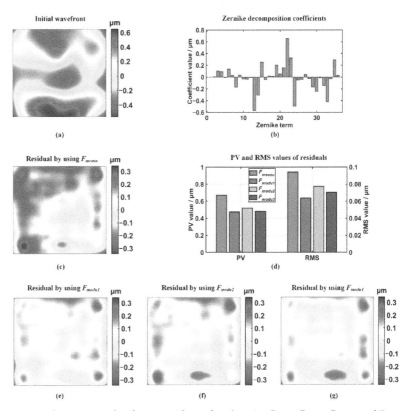

Figure 9. Correction results of a measured wavefront by using F_{measu}, F_{modu1}, F_{modu2}, and F_{modu3}. (**a**) Initial wavefront. (**b**) Zernike decomposition coefficients of the initial wavefront. (**c**) Residual by using F_{measu}. (**d**) PV and RMS values of the residual. (**e**) Residual by using F_{modu1}. (**f**) Residual by using F_{modu2}. (**g**) Residual by using F_{modu3}.

4. Experiment

An experiment is conducted to investigate the improvement of the wavefront correction capability by the IFFM under practical conditions. The configuration of the experimental setup is displayed in Figure 10.

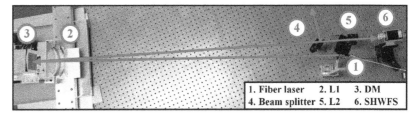

Figure 10. Configuration of the experimental setup.

The laser beam emitted from a fiber laser (Zweda Technology, 1053 nm, 0.14 numerical aperture fiber) first travels through lens L1 (1.5 m focal length) and then is reflected by the lab-manufactured DM (84 × 84 mm size, 49 square-distribution actuators) to a beam splitter where it is divided into two beams. One beam is reflected by the splitter into the rest of the optical path, and the other travels through the splitter and is collimated by lens L2 (100 mm focal length). The collimated beam enters into the SHWFS and its wavefront aberration is measured. The SHWFS consists of a micro-lens array (20 × 20 sub-apertures, 300 μm sub-aperture interval, 6 × 6 mm size) and a charge-coupled device (Basler's piA1000-48gm, 1004 × 1004 pixels, 7.4 × 7.4 μm pixel size and 7.4 × 7.4 mm sensor size). A personal computer (PC) is employed for data processing and control signal calculation, and a high-voltage driver is employed to apply the control signal to the DM.

The wavefront correction process of the AO system in the experiment includes four steps as illustrated in Figure 11: the IF measurement, the IF filtering, the wavefront measurement and the closed-loop correction. The IF of the DM is measured by using the SHWFS and the obtained data are transmitted to the PC. The IFFM is operated to modulate the measured IF F_{measu} into three filtered IFs (F_{modu1}, F_{modu2} and F_{modu3}). The initial wavefront w of the laser beam is measured by the SHWFS. The control signal u is calculated and applied on the DM to generate the compensating wavefront. To compare the wavefront correction capability between the measured IF F_{measu} and the filtered IFs (F_{modu1}, F_{modu2} and F_{modu3}), in the experiment, the wavefront correction is carried out both without and with the IF filtering, which is listed as filtering off/on in Figure 11.

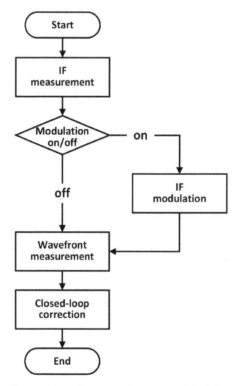

Figure 11. Wavefront correction process of the AO system.

The IF filtering results are shown in Figure 12. Note that unlike the simulation, in the experiment only the actual IF F_{measu} could be measured while the ideal IF F_a of the DM and the measurement noise δ could not be obtained. According to Figure 12, the measurement noise exists and the measured IF $F_{measu}(25)$ has random fluctuations on the whole surface shape. After filtering by using the IFFM, the measurement noises are depressed in $F_{modu1}(25)$, $F_{modu2}(25)$ and $F_{modu3}(25)$. Based on the measured IF and the filtered IF, the wavefront correction is carried out and the results are illustrated in Figure 13.

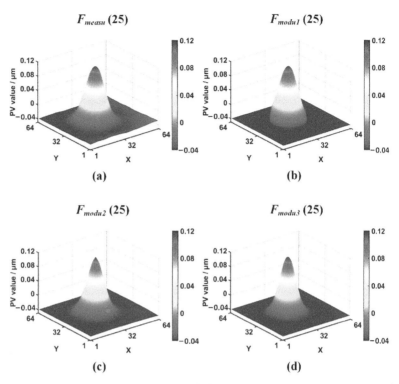

Figure 12. From the measured IF (**a**) F_{measu}, the filtered IFs (**b**) F_{modu1}, (**c**) F_{modu2}, (**d**) F_{modu3} for the 25th actuator are achieved.

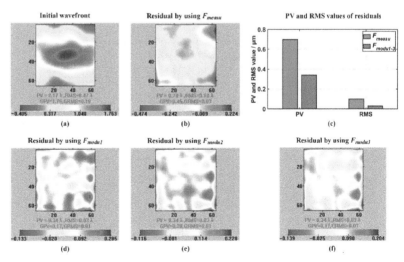

Figure 13. Correction results by using the measured IF F_{measu} and three filtered IFs $F_{modu1-3}$. (**a**) Initial wavefront. (**b**) Residual by using F_{measu}. (**c**) PV and RMS values of residuals. (**d**) Residual by using F_{modu1}. (**e**) Residual by using F_{modu2}. (**f**) Residual by using F_{modu3}.

As shown in Figure 13, the initial wavefront has the PV value of 2.17 λ ($\lambda = 1053$ nm) and the RMS value of 0.47 λ. After correction by F_{measu}, the initial aberration is depressed and the residual wavefront r has the PV value of 0.70 λ and the RMS value of 0.10 λ. It could be seen from Figure 13b that there still exists uncorrected aberrations (e.g., a distinct local convex on the right top of the residual wavefront) in the residual wavefront corrected by F_{measu}. The possible causes of the uncorrected aberrations are the measurement noise, the correction algorithm, the system structure, the DM structure and the linearity error of the correction. Compared with F_{measu}, the correction results by using F_{modu1}, F_{modu2} and F_{modu3} are improved with a 51% reduction in PV value and a 97% reduction in RMS value, while no distinct uncorrected aberrations occur in the residual wavefronts. Thus, the residual wavefronts corrected by the filtered three IFs are much smoother than those corrected by the measured IF, as can be seen in Figure 13. From Figure 13d–f it could be seen that the three residuals differ from each other in surface shape distribution. In the upper-right area of the residuals, the largest local wavefront variation comes from the residual corrected by F_{modu1}, while the smallest comes from F_{modu3}. In the central area, the local wavefront variation of the residuals corrected by F_{modu1} and F_{modu3} are both smaller than that by F_{modu2}. Nevertheless, it could also be seen that the PV and RMS values of the closed-loop residuals corrected by the three filtered IFs $F_{modu1-3}$ are very close, with differences both less than 0.1 λ. In terms of the closed-loop correction capability, the three filtered IFs are close to each other. In conclusion, experimental results indicate that the wavefront correction results of F_{modu1}, F_{modu2} and F_{modu3} are better than F_{measu}, and that the IFFM could effectively improve the wavefront correction capability of the practical AO system.

5. Discussion

In this paper, we propose the IFFM to improve the closed-loop correction capability of an AO system. In a practical wavefront sensor AO system, the measured IF and wavefront are used to calculate the control signals of the DM. As we know, the measurement is nonideal and the measurement noises always exist, which would result in the difference between the measured IF and the ideal IF. This difference finally leads to the failure to calculate the optimal control signal during each iteration of the closed-loop correction, and thus the correction ability of the AO system is degraded. The proposed IFFM depresses the measurement noises by filtering the measured IF into a new IF for the calculation of the control signals. With the filtered IF, the control signal of the closed-loop correction could converge to the optimal one, which improves the correction capability of the AO system. Compared with the measurement accuracy improvement methods proposed in [26–28], which require multiple IF measurements in the closed-loop correction process, the IFFM requires only a single IF measurement in the precorrection process, making the overall correction process more controllable. Meanwhile, the high precision IF measurement and fitting methods proposed in [29–31] could also help improve the closed-loop correction ability, while these methods are implemented with additional high precision surface measuring equipment. Since the IFFM is operated on a software platform without the need of additional measuring equipment, compared with these methods proposed in [29–31], the IFFM has better applicability and flexibility as well as practicality.

Nonetheless, the IFFM is inapplicable to the wavefront sensorless AO system where neither IF measurement nor wavefront measurement is taken for closed-loop correction. For example, the closed-loop correction mentioned in [19,20] takes the far-field beam quality as the evaluation factor without the IF measurement.

It should also be noticed that the IFFM is implemented on the wavefront controller during the precorrection process. For an AO system, the implementation of the IFFM requires no additional hardware and does not affect the correction speed. At the same time, the calculation of the IFFM is relatively simple and does not need to occupy a lot of computing resources of the wavefront controller. Finally, our method provides an approach with good applicability and practicability to improve the closed-loop correction capability of an AO system in the case of measurement noise. As can be seen from the simulation and experimental results, all the three filtered IFs we present could be adopted to reduce the measurement noise effectively. The difference between the three methods is small and could be chosen according to experiment conditions and needs.

6. Conclusions

This paper presents the IFFM to depress the negative effect of the IF measurement noise on the wavefront correction capability of the AO system in laser systems. The CAM is proposed to investigate the effect of the IF measurement noise and the improvement of the IFFM on the correction ability. Three types of filtering methods are devised in the IFFM to obtain the filtered IF from the measured IF. In the simulation, the Zernike mode aberrations and an actual measured wavefront are taken as the correction object. Simulation results reveal that the occurrence of measurement noise will degrade the wavefront correction capability and the filtered IF has better correction capability than the measured IF, as the filtered IFs have characteristics of less random fluctuations and are closer to the ideal IF. An experiment is carried out to investigate the improvement of the IFFM on the wavefront correction capability. Both simulation and experimental results indicate that the IFFM could depress the negative effect of the IF measurement noise and improve the wavefront correction capability effectively. Additionally, the IFFM does not require external hardware and affects the correction speed of the AO system.

Author Contributions: Conceptualization, L.H.; Data curation, Y.Z.; Formal analysis, Y.Z.; Funding acquisition, L.H.; Investigation, Y.Z.; Methodology, Y.Z.; Project administration, L.H.; Resources, M.L. and L.H.; Software, Y.Z. and S.L.; Supervision, L.H.; Validation, D.W. and Q.X.; Visualization, Y.Z.; Writing—original draft, Y.Z.; Writing—review & editing, Y.Z. and L.H. All authors have read and agreed to the published version of the manuscript.

Funding: This research was funded by Science and Technology on Plasma Physics Laboratory, China Academy of Engineering Physics, grant number 6142A04180304 and National Natural Science Foundation of China, grant number 61775112.

Data Availability Statement: The data presented in this study are available on request from the corresponding author.

Acknowledgments: Than authors are grateful to other colleagues for their help during the period of experimental measurement.

Conflicts of Interest: The authors declare no conflict of interest.

References

1. Zhang, X.; Arcidiacono, C.; Conrad, A.R.; Herbst, T.M.; Gaessler, W.; Bertram, T.; Ragazzoni, T.; Schreber, L.; Diolaiti, E.; Kuerster, M.; et al. Calibrating the interaction matrix for the LINC-NIRVANA high layer wavefront sensor. *Opt. Express* **2012**, *20*, 8078–8092. [CrossRef] [PubMed]
2. Lardiere, O.; Conan, R.; Bradley, C.; Jackson, K.; Herriot, G. A laser guide star wavefront sensor bench demonstrator for TMT. *Opt. Express* **2008**, *16*, 5527–5543. [CrossRef] [PubMed]
3. Baranec, C.; Lloyd -Hart, M.; Milton, N.M.; Stalcup, T.; Snyder, M.; Vaitheeswaran, V.; McCarthy, D.; Angel, R. Astronomical imaging using ground-layer adaptive optics. *Proc. SPIE* **2007**, *6691*, 66910N. [CrossRef]
4. Van Dam, M.A.; Bouchez, A.H.; Le Mignant, D.; Johansson, E.M.; Wizinowich, P.L.; Campbell, R.D.; Chin, J.C.Y.; Hartman, S.K.; Lafon, R.E.; Stomski, J.P.J.; et al. The W. M. Keck Observatory Laser Guide Star Adaptive Optics System: Performance Characterization. *Publ. Astron. Soc. Pac.* **2006**, *118*, 310–318. [CrossRef]
5. Esposito, S.; Riccardi, A.; Fini, L.; Puglisi, A.T.; Pinna, E.; Xompero, M.; Briguglio, R.; Quirós-Pacheco, F.; Stefanini, P.; Guerra, J.C.; et al. First light AO (FLAO) system for LBT: Final integration, acceptance test in Europe, and preliminary on-sky commissioning results. *Proc. SPIE* **2010**, *7736*, 773609. [CrossRef]
6. Martinache, F.; Guyon, O.; Clergeon, C.; Garrel, V.; Blain, C. The Subaru coronagraphic extreme AO project: First observations. *Proc. SPIE* **2012**, *8447*, 84471Y. [CrossRef]
7. Neichel, B.; Rigaut, F.; Vidal, F.; van Dam, M.A.; Garrel, V.; Carrasco, E.R.; Pessev, P.; Winge, C.; Boccas, M.; d'Orgeville, C.; et al. Gemini multiconjugate adaptive optics system review—II. Commissioning, operation and overall performance. *Mon. Not. R. Astron. Soc.* **2014**, *440*, 1002–1019. [CrossRef]

8. Sauvage, J.; Fusco, T.; Petit, C.; Costille, A.; Mouillet, D.; Beuzit, J.; Dohlen, K.; Kasper, M.; Suarez, M.; Soenke, C.; et al. SAXO: The extreme adaptive optics system ofSPHERE (I) system overview and global laboratory performance. *J. Astron. Telesc. Instrum. Syst.* **2016**, *2*, 025003. [CrossRef]
9. Rahman, S.A.; Booth, M.J. Direct wavefront sensing in adaptive optical microscopy using backscattered light. *Appl. Opt.* **2013**, *52*, 5523–5532. [CrossRef]
10. Bonora, S.; Jian, Y.F.; Zhang, P.F.; Zam, A.; Pugh, E.N.; Zawadzki, R.J.; Sarunic, M.V. Wavefront correction and high-resolution in vivo OCT imaging with an objective integrated multi-actuator adaptive lens. *Opt. Express* **2015**, *23*, 21931–21941. [CrossRef]
11. Sahu, P.; Mazumder, N. Advances in adaptive optics–based two-photon fluorescence microscopy for brain imaging. *Lasers Med. Sci.* **2020**, *35*, 317–328. [CrossRef] [PubMed]
12. Zacharias, R.A.; Beer, N.R.; Bliss, E.S.; Burkhart, S.C.; Cohen, S.J.; Sutton, S.B.; Van Atta, R.L.; Winters, S.E.; Salmon, J.T.; Stolz, C.J.; et al. National Ignition Facility alignment and wavefront control. *Proc. SPIE* **2004**, *5341*, 168–179. [CrossRef]
13. Sacks, R.; Auerbach, J.; Bliss, E.; Henesian, M.; Lawson, J.; Manes, K.; Renard, P.; Salmon, T.; Trenholme, J.; Williams, W.; et al. Application of adaptive optics for controlling the NIF laser performance and spot size. *Proc. SPIE* **1999**, *3492*, 344–354. [CrossRef]
14. Van Wonterghem, B.M.; Murray, J.R.; Campbell, J.H.; Speck, D.R.; Barker, C.E.; Smith, I.C.; Browning, D.F.; Behrendt, W.C. Performance of a prototype for a large-aperture multipass Nd:glass laser for inertial confinement fusion. *Appl. Opt.* **1997**, *36*, 4932–4953. [CrossRef]
15. Hammer, D.X.; Ferguson, R.D.; Bigelow, C.E.; Iftimia, N.V.; Ustun, T.E.; Burns, S.A. Adaptive optics scanning laser ophthalmoscope for stabilized retinal imaging. *Opt. Express* **2006**, *14*, 3354–3367. [CrossRef]
16. Fernandez, E.J.; Vabre, L.; Hermann, B.; Unterhuber, A.; Povazay, B.; Drexler, W. Adaptive optics with a magnetic deformable mirror: Applications in the human eye. *Opt. Express* **2006**, *14*, 8900–8917. [CrossRef]
17. Marcos, S.; Werner, J.S.; Burns, S.A.; Merigan, W.H.; Artal, P.; Atchison, D.A.; Hampson, K.M.; Legras, R.; Lundstrom, L.; Yoon, G.; et al. Vision science and adaptive optics, the state of the field. *Vision Res.* **2017**, *132*, 3–33. [CrossRef]
18. Goodno, G.D.; Komine, H.; McNaught, S.J.; Weiss, S.B.; Redmond, S.; Long, W.; Simpson, R.; Cheung, E.C.; Howland, D.; Epp, P.; et al. Coherent combination of high-power, zigzag slab lasers. *Opt. Lett.* **2006**, *31*, 1247–1249. [CrossRef]
19. Sun, L.C.; Guo, Y.D.; Shao, C.F.; Li, Y.; Zheng, Y.M.; Sun, C.; Wang, X.J.; Huang, L. 10.8 kW, 2.6 times diffraction limited laser based on a continuous wave Nd:YAG oscillator and an extra-cavity adaptive optics system. *Opt. Lett.* **2018**, *43*, 4160–4163. [CrossRef]
20. Xu, L.; Wu, Y.C.; Du, Y.L.; Wang, D.; An, X.C.; Li, M.; Zhou, T.J.; Shang, J.L.; Wang, J.T.; Liu, Z.W.; et al. High brightness laser based on Yb:YAG MOPA chain and adaptive optics system at room temperature. *Opt. Express* **2018**, *26*, 14592–14600. [CrossRef]
21. Brousseau, D.; Borra, E.F.; Thibault, S. Wavefront correction with a 37-actuator ferrofluid deformable mirror. *Opt. Express* **2007**, *15*, 18190–18199. [CrossRef]
22. Bonora, S. Distributed actuators deformable mirror for adaptive optics. *Opt. Commun.* **2011**, *284*, 3467–3473. [CrossRef]
23. Gonzalez-Nunez, H.; Bechet, C.; Ayancan, B.; Neichel, B.; Guesalaga, A. Effect of the influence function of deformable mirrors on laser beam shaping. *Appl. Opt.* **2017**, *56*, 1637–1646. [CrossRef]
24. Fernandez, E.J.; Artal, P. Membrane deformable mirror for adaptive optics: Performance limits in visual optics. *Opt. Express* **2003**, *11*, 1056–1069. [CrossRef] [PubMed]
25. Mansell, J.; Jameson, J.; Henderson, B. Advanced deformable mirrors for high-power lasers. *Proc. SPIE* **2014**, *9083*, 90830O. [CrossRef]
26. Zou, W.Y.; Burns, S.A. High-accuracy wavefront control for retinal imaging with Adaptive-Influence-Matrix Adaptive Optics. *Opt. Express* **2009**, *17*, 20167–20177. [CrossRef] [PubMed]
27. Huang, L.; Ma, X.K.; Bian, Q.; Li, T.H.; Zhou, C.L.; Gong, M.L. High-precision system identification method for a deformable mirror in wavefront control. *Appl. Opt.* **2015**, *54*, 4313–4317. [CrossRef]
28. Ren, Z.L.; Liu, J.; Zhang, Z.T.; Chen, Z.T.; Liang, Y.H. High-resolution wavefront reconstruction using multiframe Shack–Hartmann wavefront sensor measurements. *Opt. Eng.* **2020**, *59*, 113102. [CrossRef]
29. Zheng, Y.M.; Sun, C.; Dai, W.J.; Zeng, F.; Xue, Q.; Wang, D.E.; Zhao, W.C.; Huang, L. High precision wavefront correction using an influence function optimization method based on a hybrid adaptive optics system. *Opt. Express* **2019**, *27*, 34937–34951. [CrossRef]
30. Huang, L.H.; Rao, C.H.; Jiang, W.H. Modified Gaussian influence function of deformable mirror actuators. *Opt. Express* **2008**, *16*, 108–114. [CrossRef]
31. Roopashree, M.B.; Vyas, A.; Prasad, B.R. Influence Function Measurement of Continuous Membrane Deformable Mirror Actuators Using Shack Hartmann Sensor. *AIP Conf. Proc.* **2011**, *1391*, 453–455. [CrossRef]
32. Rosen, S.; Eldert, C. Least-Squares Method for Optical Correction. *J. Opt. Soc. Am.* **1954**, *44*, 250–252. [CrossRef]
33. Claflin, E.S.; Bareket, N. Configuring an electrostatic membrane mirror by least-squares fitting with analytically derived influence functions. *J. Opt. Soc. Am.* **1986**, *3*, 1833–1839. [CrossRef]
34. Feng, Z.X.; Huang, L. Over compensation algorithm for laser beam shaping using a deformable freeform mirror. *Optik* **2019**, *198*, 163250. [CrossRef]
35. Tabatabaian, M. *COMSOL for Engineers*; Mercury Learning & Information: Herndon, VA, USA, 2014.
36. Higham, D.J.; Higham, N.J. *MATLAB Guide*, 3rd ed.; Siam: Philadelphia, PA, USA, 2016.
37. Wang, J.Y.; Silva, D.E. Wave-front interpretation with Zernike polynomials. *Appl. Opt.* **1980**, *19*, 1510–1518. [CrossRef] [PubMed]
38. Alda, J.; Boreman, G.D. Zernike-based matrix model of deformable mirrors: Optimization of aperture size. *Appl. Opt.* **1993**, *32*, 2431–2438. [CrossRef]

Communication

State-of-the-Art Technologies in Piezoelectric Deformable Mirror Design

Vladimir Toporovsky [1,*], Alexis Kudryashov [1], Arkadiy Skvortsov [2], Alexey Rukosuev [1], Vadim Samarkin [1] and Ilya Galaktionov [1]

[1] Sadovsky Institute of Geosphere Dynamics of Russian Academy of Sciences, Leninskiy pr. 38/1, 119334 Moscow, Russia; kud@activeoptics.ru (A.K.); alru@nightn.ru (A.R.); samarkin@nightn.ru (V.S.); galaktionov@activeoptics.ru (I.G.)

[2] Transport Faculty Department, Moscow Polytechnic University, Bolshaya Semenovskaya Str. 38, 107023 Moscow, Russia; a.a.skvortsov@mospolytech.ru

* Correspondence: topor@activeoptics.ru; Tel.: +7-985-196-37-33

Abstract: In this work, two advanced technologies were applied for manufacturing a bimorph wavefront corrector: laser ablation, to vaporize conductive silver coating from piezoceramic surface, and parallel-gap resistance microwelding, to provide a reliable electrical contact between the piezodisk surface silver electrodes and copper wires. A step-by-step guide for bimorph mirror production is presented, together with the 'bottlenecks'. Optimization of the laser ablation technique was carried out using an Nd:YAG laser with an output power of 4 W and a frequency of 20 kHz. A comparison of the ultrasonic welding and parallel-gap resistance microwelding methods was performed. The tensile strength in the first case was in the range of 0.2 ... 0.25 N for the system 'copper wire–silver coating'. The use of resistance welding made it possible to increase the value of this parameter for the same contact pair by almost two times (0.45 ... 0.5 N).

Keywords: adaptive optics; deformable mirror; laser ablation; parallel-gap resistance microwelding; wavefront control

1. Introduction

The evolution of adaptive optics is associated with the problem of developing reliable and miniature devices for the compensation of laser beam aberrations [1–3]. Wavefront correctors are an essential part of any adaptive optical system [4–6]. The main function of this element as an executive unit is to directly compensate distortions in light radiation through changes in shape by means of various mechanisms: electrostatic [7], pneumatic [8], electromagnetic [9], or piezoelectric [10,11]. Different types of correcting devices have been proposed and developed [12–16]. Moreover, there is a tendency to produce such devices with high numbers of control elements [17] and increased wavefront corrector reliability [18], while simultaneously reducing the final product price. The most suitable types of deformable mirror, due to their low cost and high reliability, are bimorph ones [19,20]. Moreover, our scientific group is widely known for the development of such wavefront correctors [21–25]. Nevertheless, there is a problem in the creation of these flexible mirrors resulting from the high spatial resolution of the control electrodes.

High-tech methods and materials are widely used for manufacturing bimorph wavefront correctors. For example, the CVD technique for the production of mirror substrates with embedded cooling channels [26] and additive technologies like laser stereolithography have been applied [27,28]. Moreover, nanotechnological methods such as magnetron sputter deposition and evaporation for the creation of reflective or conductive coatings for adaptive mirrors are widespread [29,30].

Small-aperture bimorph mirrors should be produced using innovative methods such as ultrasonic welding [31], parallel-gap contact microwelding [32], and laser ablation [33] techniques.

In the second section of this article, the key principle and the design of the bimorph deformable mirror are demonstrated, along with the main manufacturing steps. The third section presents a description of production technologies that make it possible to reduce the corrector size (mirror diameter) and to provide the stable performance of the adaptive optical element. In Section 4, the results of manufacturing a bimorph deformable mirror with a small aperture and a high density of control elements are presented.

2. Materials and Methods

Bimorph flexible mirrors consist of a polished passive substrate made of glass, silicon or copper, along with an active piezoceramic plate with the conductive layer of silver or nickel (Figure 1a). On the outer surface of the piezoplate, an electrode grid is placed (Figure 1b,c). While applying control voltages to the electrodes, the transverse piezoceramic effect leads to the expansion (or compression) of the piezoplate, causing deformation (bend) to the mirror surface. By varying the control signals on the electrodes, the desired shape of the reflective surface can be realized.

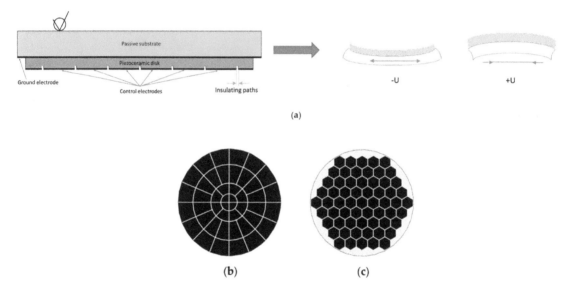

Figure 1. Traditional bimorph deformable mirror: (**a**) conventional design and principle of operation, (**b**) keystone electrode grid, (**c**) honeycomb electrode grid.

The main steps of manufacturing traditional bimorph mirrors are as follows: (1) the piezoceramic plate is glued to the passive substrate; (2) after that, the electrode grid is drawn on the piezo plate surface using the photolithographic technique; (3) the substrate is polished in order to obtain a high-quality reflecting (mirror) surface; (4) in the next step, a highly reflective coating is deposited (>99.9%); (5) by using conductive glue, the contact wires are connected to the mirror electrodes; and finally, (6) the electrical wiring and mirror assembly are placed in the housing.

This manufacturing process for bimorph correctors is necessary for mirrors with average (30–100 mm) and large (>100 mm) apertures. Furthermore, during assembly, preliminary mechanical correction is performed for large-aperture mirrors by changing the tension of the adjustment screws located in the mounting. This technical solution makes it possible to reduce the influence of the initial aberrations (astigmatism, coma) arising from assembly and deposition of the reflective coating [34].

However, in the case of probing of the atmosphere using LiDAR [35], in free space communication systems [36] or in laser materials processing [37], as well as in microscopy [38], it is necessary to use small-aperture mirrors (<30 mm). At the same time, the number of control elements should be sufficient to compensate for both low- and high-order wavefront aberrations [39]. The deformable mirror should also provide a reasonable stroke of the reflective surface in order to compensate for large-scale wavefront aberrations. Increasing the local deflection of the mirror can be achieved by using a thin passive substrate, but its thickness must be sufficient to avoid the print-through effect of the control electrode grid [40].

Piezoceramic is a porous material, which leads to some inconveniences with respect to placing the electrode grid on its surface. The traditional technique of chemically etching the insulating paths in the conductive layers deposited on the piezodisk is not quite applicable in this case. Nevertheless, this process has proven to be an ideal tool for the creation of dielectric paths in bimorph deformable mirrors, and can be applied for average- and large-aperture mirrors [41]. Therefore, to avoid short circuit between neighbor electrodes, it is necessary to make rather wide paths. The particles of the conductive material (silver or nickel) might penetrate inside the ceramic hollows, thus creating unnecessary electrical connection between electrodes. At the same time, different methods for creating insulating paths (dielectric tracks) on the disk surface using laser the ablation technique can lead to the unexpected removal of piezoceramic material. This might destroy the whole piezodisk, and as a result introduce the so-called print-through effect on the mirror surface [30].

Additionally, the wiring of the electric contacts is an important problem arising during the manufacturing of small-aperture deformable mirrors because of the reduced size of both the electrodes and the contact area (to 5–12 mm^2). For such an electrode size, the traditional gluing methods with the conductive epoxy are not suitable. The use of conventional soldering methods is impossible as well, considering that the heating of the piezodisk will result in the local depolarization of the piezoceramic, resulting in the failure of the mirror.

For this reason, to overcome these problems, we propose the use of laser ablation and microwelding techniques for the development of bimorph wavefront correctors.

Laser ablation methods are widely used for the synthesis of colloidal nanoparticles in liquids [42] and nanostructural coatings [43]. Additionally, this technique can be applied for the removal of layers from various surfaces of materials [44]. The working principle is as follows: the light from the emitter passes through a set of mirrors to the objective lens and is focused on the object surface. During the processing, the upper layers of the treated material evaporate, and after that, insulating paths between electrodes appear. For the treatment of different metals, lasers with the wavelength of about 1 μm are applied [45], such as Nd:YAG-lasers. Additionally, lasers with a second-harmonic generation mode can often be used.

We exploited a laser ablation setup with a Q-switched Nd:YAG laser with a pulse energy of 0.67 mJ, a pulse duration of 100 ns, and a frequency repetition rate in the range from 20 to 200 kHz. Average laser radiation power was up to 20 W. The minimal focal spot on the object surface was about 50 μm. Laser ablation technology possesses the favorable characteristics of high processing speed and precise materials treatment. Additionally, the full automatization of the process guarantees the high repeatability of the results unlike the chemical etching. Nevertheless, the use of such a technique is restricted by the non-uniformity of the layers of the workpieces [46]. As was mentioned earlier, the overheating of the bimorph plate can result in additional initial distortions of the reflective surface of the mirror and decrease the efficiency of correction of wavefront aberrations. Obviously, depending on the piezoplate properties (conductive layer thickness, material of coating), the parameters of the laser beam for electrode coating removal should be adjusted.

In our case, we used piezoceramic disks with a thickness of 0.2 mm. Their surface was coated with a conductive silver layer with a thickness of about 20 μm. In the setup for creating insulating paths, a radiation with an average power of 4 W with a pulse frequency of 20 kHz was used. The width of the dielectric tracks was chosen as 300 µm to avoid electrical breakdown between electrodes during mirror operation.

Using these ablation parameters, we manufactured a piezoplate with 37 electrodes in a keystone geometry on a 30 mm aperture (Figure 2). Thus, the average area of one electrode was equal to 10 mm^2. An alternative method to gluing techniques for wiring electrical contacts for this bonding area, in our opinion, is microwelding approaches (thermocompression, ultrasonic, parallel-gap resistance microwelding), which are widely used in contemporary microelectronics.

Figure 2. The electrode grid placed on the piezodisk surface.

Pad-to-pad wiring is one of the most widely used approaches for bonding electrotechnical elements [47]. Traditional welding methods are almost never used, because of the small sizes of the contact areas. Microcontacts are realized, as a rule, by means of gold, aluminum, or copper 30–150 µm wires. In our work, we chose three widely used microwelding methods—thermocompression, ultrasonic, parallel-gap resistance. We analyzed the most suitable welding techniques for the design of bimorph deformable mirrors.

Thermocompression welding is used to attach the pins to the semiconductor crystals, for welding the microwires, for welding the flat cable cores to connector pins, etc. [48]. However, the local heating of the processing materials can result in temperatures of up to 650 °C. Such high thermal stresses can damage the structure of the welding components, meaning that this technique cannot be used for the manufacture of piezoelectric deformable mirrors.

To overcome this disadvantage the ultrasonic approach of microcontacting was proposed. This method is characterized by the application of ultrasonic energy to the contact area [49]. This results in bursting and burning of the solid fatty films in the region of the connection point, the plastic deformation of the material and intensive diffusion. The high-frequency vibrations lead to a reliable connection of the workpieces by means of the plastic deformations of the welding parts. We carried out experiments in which aluminum and copper wires (with the diameter of 50 . . . 100 µm) were welded to various surfaces (Al, Cu, Ag).

However, it should be mentioned that, together with the minimal heating stresses, we were not able to achieve a significant value for tensile strength P. For the contact system 'aluminum wire (diameter d = 80 µm)–silver film (thickness h = 20 µm)' the tensile strength value with ultrasonic microwelding was $P_{Al(US)}$ = 0.15 . . . 0.20 N, and for the system 'copper wire (d = 80 µm)–silver film (h = 20 µm)' the tensile strength value increased to $P_{Cu(US)}$ = 0.2 . . . 0.25 N.

Greater values of P_{Cu} were obtained when using the parallel-gap microwelding technique. This approach is a variation of the resistance welding method [32] adapted for the design features of deformable piezoelectric mirrors (Figure 3a). The welding materials in this case were copper, aluminum and silver. The thickness of the silver films was in the range of 15–25 µm, and the diameter of the copper and aluminum wires was 80 µm. Tungsten was used as an electrode material.

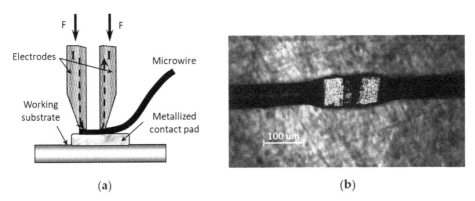

Figure 3. Parallel-gap resistance microwelding: (**a**) scheme of setup, (**b**) result of welding of copper wire to silver coating on piezodisk.

Before welding, the workpieces were annealed to decrease their inner stresses and increase their plasticity. The surfaces of the materials were also degreased with chemical solvents. The welding process was performed with an electrode made of tungsten as the current-conductive parts, with a gap of 0.02 ... 0.25 mm depending on the thickness and diameter of the workpieces.

The efficiency of parallel-gap resistance microwelding was estimated by analysis of the ensile strength value. For contact pair 'copper wire (d = 80 µm)–silver film (h = 20 µm)', this parameter was $P_{Cu(R)}$ = 0.45 ... 0.50 N, depending on the various welding regimes.

In Figure 3b, the results of welding copper wire with a diameter of 80 µm to silver film with a thickness of 20 µm on the piezoceramic plate as part of the bimorph deformable mirror are shown.

3. Results and Discussion

The optimization of the traditional laser ablation and parallel-gap resistance microwelding techniques for adaptive optics tasks made it possible to manufacture a 37-electrode bimorph deformable mirror (Figure 4a,b) with an active diameter of 30 mm.

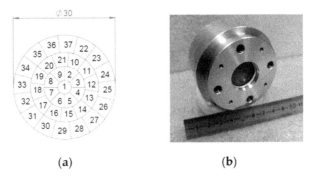

Figure 4. Developed bimorph deformable mirror: (**a**) electrode grid, (**b**) assembled deformable mirror.

After assembling the bimorph deformable mirror, we investigated its main parameters—the initial quality of the reflective surface (as the most important characteristic), local and total stroke of the mirror, and the first resonant frequency.

The initial surface quality was measured using a Fizeau interferometer Zygo Mark GPI-XPS. As seen in Figure 5a, the print-through effect was not visible on the reflective surface of the mirror. The initial surface distortions were equal to PV = 1.72 µm, RMS = 0.46 µm, when taking into account the total surface curvature (Figure 5b), and PV = 0.42 µm, RMS = 0.08 µm without defocus aberration (Figure 5c). The presence of astigmatism aberration is intrinsic to the initial surface because of the fixing mechanism of the bimorph wavefront corrector in the housing.

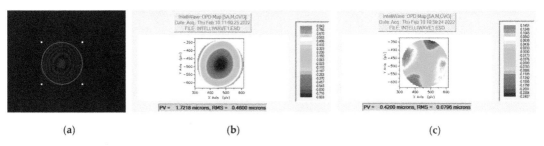

(a) (b) (c)

Figure 5. Initial shape of the developed deformable mirror: (**a**) image as interferogram, (**b**) initial surface quality with the total curvature (PV = 1.72 µm, RMS = 0.46 µm), (**c**) initial surface without defocus aberration (PV = 0.42 µm, RMS = 0.08 µm).

To measure the stroke of the mirror, we used a diagnostic setup based on a Shack-Hartmann wavefront sensor, simulating the conventional closed-loop adaptive optical system (Figure 6). The setup consisted of a diode laser with an average output power of 5 mW at a wavelength of 650 nm, a lens with a focal length of 500 mm, the bimorph deformable mirror, and a Shack–Hartmann wavefront sensor [50]. The wavefront sensor included a CMOS camera, a lenslet with the microlens focal length of 5 mm and a pitch of 120 µm, and a 6× beam-reducing telescope. To adjust the optical system, a tracer laser with a narrow parallel beam with a wavelength of 650 nm was used. The radiation of this laser was propagated through the centers of the optical elements. The diagnostic and tracer laser beams were combined on the diaphragm.

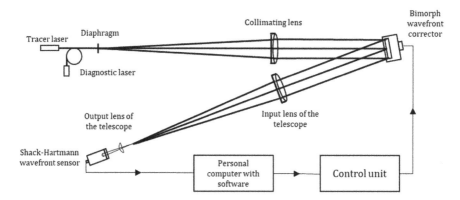

Figure 6. Diagnostic setup for investigation of the parameters of the deformable mirror.

It is obvious that the local stroke of the inner electrodes will be smaller than that of electrodes at the periphery of the mirror. Therefore, we measured the value of this parameter for two different electrodes—#7 and #32. The stroke was found to be 1.4 µm (PV) and 4.7 µm (PV) for the 7th and 32nd electrodes, respectively. These values are comparable to those of bimorph deformable mirrors with significantly larger electrode areas [51].

Additionally, we measured the total stroke of the developed deformable mirror. We applied the maximum control voltage to all mirror electrodes in order to achieve formation of defocus aberration. The largest value was equal to 12 µm (PV).

One of the significant parameters in deformable mirror operation in closed-loop adaptive optical systems is speed—how fast the mirror can introduce different wavefront aberrations. This parameter is directly related to the first resonant frequency of the wavefront corrector. For bimorph mirrors, it can be estimated using the following formula [52]:

$$f = K \frac{1}{2\pi} \sqrt{\frac{Dg}{wr^4}},$$

where K—scaling coefficient depending on the boundary conditions (shown in Table 1) [53], D—effective elastic constant for the plate, g—gravitational acceleration, w—loading per unit area of the plate, including its own weight ($w = g\sum \rho_k t_k$, ρ_k and t_k correspond to the density and thickness of each layer).

Table 1. Values of the scaling constant K for different boundary conditions.

Boundary Conditions	K
Simply constrained	4.99
Edge clamped	10.2
Edge free	5.25

Thus, the estimated value of the first resonant frequency for the bimorph mirror with an overall thickness of 1.6 mm (0.2 mm for piezoceramic and 1.4 mm for glass substrate) should be equal to 13.4 kHz. To measure the real value of this parameter, we used an oscilloscope Tektronix DPO-2004B and a sine wave generator. A sine voltage with a frequency range from 0.1 to 18 kHz was applied to electrode #1. Due to the piezoelectric effect, AC voltage was detected on the neighboring electrode #2 and was registered using the oscilloscope [54]. The amplitude of the output signal was stable up to a frequency of 12 kHz, and the first resonance was found at a frequency of 13.2 kHz (Figure 7). More detailed analysis of the temporal behavior using the Bode curve [55] shows the possibility of using such a deformable mirror in a closed-loop adaptive system working with a frequency (in terms of frames per second) of 10 kHz.

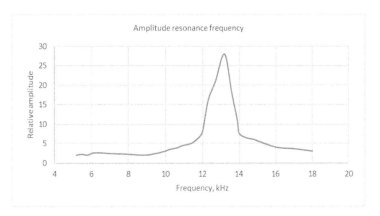

Figure 7. Amplitude–frequency response of the bimorph wavefront corrector.

4. Conclusions

We developed a 37-electrode bimorph deformable mirror with a diameter of 30 mm. The results of using the laser ablation technique were demonstrated. This technique was proposed for removing the silver coating from the piezodisk surface. Using the Q-switched Nd:YAG laser with an average power of 4 W and a repetition rate of 20 kHz made it possible to obtain insulating paths with a width of 300 µm. It was shown that by using the ultrasonic welding approach for bonding the copper wires to the silver contact pad of the piezodisk, it was possible to achieve a tensile strength value in the range of $P_{Cu(US)}$ 0.2 ... 0.25 N, while the parallel-gap resistance microwelding technique made it possible to increase this value by two times ($P_{Cu(R)}$ = 0.45 ... 0.5 N). The stroke of the single electrode of the mirror varied from 1.4 µm to 4.7 µm, depending on electrode position. The maximum total stroke of the bimorph corrector was equal to 12 µm. The first resonant frequency of the mirror was found at 13.2 kHz. Such a deformable mirror would be useful for the compensation of low-order and high-order aberrations caused by atmospheric turbulence.

Author Contributions: Conceptualization, A.K. and A.S.; methodology, A.S.; software, I.G.; validation, V.T., V.S. and A.R.; formal analysis, V.T. and V.S.; investigation, V.T. and A.R.; resources, A.K.; data curation, V.T. and V.S.; writing—original draft preparation, V.T. and A.K.; writing—review and editing, V.T. and A.K.; visualization, V.T. and I.G.; supervision, V.S.; project administration, A.K.; funding acquisition, A.K. All authors have read and agreed to the published version of the manuscript.

Funding: The investigations of the mirror manufacturing methods (Chapter 2) were supported by the Russian Science Foundation under grant #19-19-00706P, and the investigation of the mirror properties (Chapter 3) was supported by the state assignment of the Ministry of Science and Higher Education of the Russian Federation (theme No122032900183-1).

Institutional Review Board Statement: Not applicable.

Informed Consent Statement: Not applicable.

Data Availability Statement: The data that supports the findings of this study are available from the corresponding author upon reasonable request.

Conflicts of Interest: The authors declare no conflict of interest.

References

1. Bendek, E.; Lynch, D.; Pluzhnik, E.; Belikov, R.; Klamm, B.; Hyde, E.; Mumm, K. Development of a miniaturized deformable mirror controller. *Proc. SPIE* **2016**, *9909*, 990984. [CrossRef]
2. Divoux, C.; Cugat, O.; Basrour, S.; Mounaix, P.; Kern, P.Y.; Boussey-Said, J. Miniaturized deformable magnetic mirror for adaptive optics. *Proc. SPIE* **1998**, *3353*, 850–857. [CrossRef]
3. Toporovsky, V.; Kudryashov, A.; Samarkin, V.; Panich, A.; Sokallo, A.; Malykhin, A.; Skrylev, A.; Sheldakova, J.V. Small-aperture stacked-array deformable mirror made of the piezoceramic combs. *Proc. SPIE* **2021**, *11672*, 1167215. [CrossRef]
4. Sato, T.; Ishida, H.; Ikeda, O. Adaptive PVDF piezoelectric deformable mirror system. *Appl. Opt.* **1980**, *19*, 1430–1434. [CrossRef]
5. Wei, K.; Zhang, X.; Xian, H.; Ma, W.; Zhang, A.; Zhou, L.; Guan, C.; Li, M.; Chen, D.; Chen, S.; et al. First light on the 127-element adaptive optical system for 1.8-m telescope. *Chin. Opt. Lett.* **2010**, *8*, 1019–1021. [CrossRef]
6. Kudryashov, A.; Rukosuev, A.; Nikitin, A.; Galaktionov, I.; Sheldakova, J. Real-time 1.5 kHz adaptive optical system to correct for atmospheric turbulence. *Opt. Express* **2020**, *28*, 37546–37552. [CrossRef]
7. Bifano, T. MEMS deformable mirrors. *Nat. Photonics* **2021**, *5*, 21–23. [CrossRef]
8. Moghimi, M.J.; Wilson, C.; Dickensheets, D.L. Electrostatic-pneumatic MEMS deformable mirror for focus control. In Proceedings of the 2012 International Conference on Optical MEMS and Nanophotonics, Banff, AB, Canada, 6–9 August 2012; pp. 132–133.
9. Biasi, R.; Gallieni, D.; Salinari, P.; Riccardi, A.; Mantegazza, P. Contactless thin adaptive mirror technology: Past, present, and future. *Proc. SPIE* **2010**, *7736*, 77362B. [CrossRef]
10. Toporovsky, V.V.; Kudryashov, A.V.; Samarkin, V.V.; Rukosuev, A.L.; Nikitin, A.N.; Sheldakova, Y.V.; Otrubyannikova, O.V. Cooled Stacked-Actuator Deformable Mirror for Compensation for Phase Fluctuations in a Turbulent Atmosphere. *Atmospheric Ocean. Opt.* **2020**, *33*, 584–590. [CrossRef]
11. Samarkin, V.; Aleksandrov, A.G.; Jitsuno, T.; Romanov, P.N.; Rukosuev, A.L.; Kudryashov, A. Study of a wide-aperture combined deformable mirror for high-power pulsed phosphate glass lasers. *Quantum Electron.* **2015**, *45*, 1086–1087. [CrossRef]
12. Lefaudeux, N.; Levecq, X.; Dovillaire, G.; Ballesta, J.; Lavergne, E.; Sauvageot, P.; Escolano, L. Development of a new tech-nology of deformable mirror for ultra-intense laser applications. *Nucl. Instrum. Methods Phys. Res. Sect. A Accel. Spectrometers Detect. Assoc. Equip.* **2011**, *653*, 164–167. [CrossRef]
13. Salinari, P.; Del Vecchio, C.; Biliotti, V. A study of an adaptive secondary mirror. In Proceedings of the ICO-16 (International Commission for Optics) Satellite Conference on Active and adaptive optics, Garching/Munich, Germany, 2–5 August 1993; p. 247.

14. Lück, H.; Müller, K.-O.; Aufmuth, P.; Danzmann, K. Correction of wavefront distortions by means of thermally adaptive optics. *Opt. Commun.* **2000**, *175*, 275–287. [CrossRef]
15. Ivanova, N.L.; Onokhov, A.P.; Chaika, A.N.; Resnichenko, V.V.; Yeskov, D.N.; Gromadin, A.L.; Feoktistov, N.A.; Beresnev, L.A.; Pape, D.R. Liquid-crystal spatial light modulators for adaptive optics and image processing. *Proc. SPIE* **1996**, *2754*, 180–185. [CrossRef]
16. Samarkin, V.V.; Alexandrov, A.; Toporovsky, V.; Rukosuev, A.; Kudryashov, A. Water-cooled deformable mirrors for high power beam correction. *Proc. SPIE* **2021**, *11849*, 1184917. [CrossRef]
17. Riaud, P. New high-density deformable mirrors for high-contrast imaging. *Astron. Astrophys.* **2012**, *545*, A25. [CrossRef]
18. Cornelissen, S.A.; Bifano, T.G.; Bierden, P.A. MEMS deformable mirror actuators with enhanced reliability. *Proc. SPIE* **2012**, *8253*, 47–53. [CrossRef]
19. Oya, S.; Bouvier, A.; Guyon, O.; Watanabe, M.; Hayano, Y.; Takami, H.; Iye, M.; Hattori, M.; Saito, Y.; Itoh, M.; et al. Performance of the deformable mirror for Subaru LGSAO. *Proc. SPIE* **2006**, *6272*, 62724S.
20. Alaluf, D.; Bastaits, R.; Wang, K.; Horodinca, M.; Martic, G.; Mokrani, B.; Preumont, A. Unimorph mirror for adaptive optics in space telescopes. *Appl. Opt.* **2018**, *57*, 3629–3638. [CrossRef]
21. Kudryashov, A.V.; Kulakov, V.B.; Kotsuba, Y.V.; Novikova, L.V.; Panchenko, V.Y.; Samarkin, V.V. Low-cost adaptive optical devices for multipurpose applications. *Proc. SPIE* **1999**, *3688*, 469–476. [CrossRef]
22. Galaktionov, I.; Sheldakova, J.; Nikitin, A.; Samarkin, V.; Parfenov, V.; Kudryashov, A. Laser beam focusing through a moderately scattering medium using a bimorph mirror. *Opt. Express* **2020**, *28*, 38061–38075. [CrossRef]
23. Kudryashov, A.; Alexandrov, A.; Rukosuev, A.; Samarkin, V.; Galarneau, P.; Turbide, S.; Châteauneuf, F. Extremely high-power CO2 laser beam correction. *Appl. Opt.* **2015**, *54*, 4352–4358. [CrossRef] [PubMed]
24. Wattellier, B.; Fuchs, J.; Zou, J.-P.; Chanteloup, J.C.; Bandulet, H.; Michel, P.; Labaune, C.; Depierreux, S.; Kudryashov, A.; Aleksandrov, A. Generation of a single hot spot by use of a deformable mirror and study of its propagation in an underdense plasma. *J. Opt. Soc. Am. B* **2003**, *20*, 1632–1642. [CrossRef]
25. Samarkin, V.; Aleksandrov, A.; Dubikovsky, V.; Kudryashov, A. Water-cooled bimorph correctors. *Proc. SPIE* **2005**, *6018*, 60180Z. [CrossRef]
26. Ahn, K.; Rhee, H.-G.; Yang, H.-S.; Kihm, H. CVD SiC deformable mirror with monolithic cooling channels. *Opt. Express* **2018**, *26*, 9724–9739. [CrossRef]
27. Cholleti, E.R. A Review on 3D printing of piezoelectric materials. *IOP Conf. Series Mater. Sci. Eng.* **2018**, *455*, 012046. [CrossRef]
28. Hall, S.E.; Regis, J.E.; Renteria, A.; Chavez, L.A.; Delfin, L.; Vargas, S.; Haberman, M.R.; Espalin, D.; Wicker, R.; Lin, Y. Paste extrusion 3D printing and characterization of lead zirconate titanate piezoelectric ceramics. *Ceram. Int.* **2021**, *47*, 22042–22048. [CrossRef]
29. Hickey, G.; Barbee, T.; Ealey, M.; Redding, D. Actuated hybrid mirrors for space telescopes. *Proc. SPIE* **2010**, *7731*, 773120. [CrossRef]
30. Rausch, P.; Verpoort, S.; Wittrock, U. Unimorph deformable mirror for space telescopes: Design and manufacturing. *Opt. Express* **2015**, *23*, 19469–19477. [CrossRef]
31. Neppiras, E. Ultrasonic welding of metals. *Ultrasonics* **1965**, *3*, 128–135. [CrossRef]
32. Biele, L.; Schaaf, P.; Schmid, F. Specific Electrical Contact Resistance of Copper in Resistance Welding. *Phys. Status Solidi Appl. Mater. Sci.* **2021**, *218*, 2100224. [CrossRef]
33. Chichkov, B.N.; Momma, C.; Nolte, S.; Von Alvensleben, F.; Tünnermann, A. Femtosecond, picosecond and nanosecond laser ablation of solids. *Appl. Phys. A* **1996**, *63*, 109–115. [CrossRef]
34. Samarkin, V.V.; Alexandrov, A.G.; Galaktionov, I.V.; Kudryashov, A.V.; Nikitin, A.N.; Rukosuev, A.L.; Toporovsky, V.V.; Sheldakova, Y.V. Large-aperture adaptive optical system for correcting wavefront distortions of a petawatt Ti: Sapphire laser beam. *Quantum Electron.* **2022**, *52*, 187–194. [CrossRef]
35. Tan, X.-C.; Wu, Z.-C.; Liang, Z. Effect of adaptive optical system on the capability of lidar detection in atmosphere. *Proc. SPIE* **2009**, *7284*, 72840G. [CrossRef]
36. Weyrauch, T.; Vorontsov, M. Free-space laser communications with adaptive optics: Atmospheric compensation experiments. *J. Opt. Fiber Commun. Rep.* **2004**, *1*, 355–379. [CrossRef]
37. Spalding, I. Electric-discharge pumping. *High Power Laser Sci. Eng.* **1996**, *7*, 27.
38. Booth, M.J. Adaptive optical microscopy: The ongoing quest for a perfect image. *Light. Sci. Appl.* **2014**, *3*, e165. [CrossRef]
39. Noll, R.J. Zernike polynomials and atmospheric turbulence. *J. Opt. Soc. Am.* **1976**, *66*, 207–211. [CrossRef]
40. Verpoort, S.; Rausch, P.; Wittrock, U. Characterization of a miniaturized unimorph deformable mirror for high power CW-solid state lasers. *Proc. SPIE* **2012**, *8253*, 825309. [CrossRef]
41. Zhu, Z.; Li, Y.; Chen, J.; Ma, J.; Chu, J. Development of a unimorph deformable mirror with water cooling. *Opt. Express* **2017**, *25*, 29916. [CrossRef]
42. Arakelyan, S.M.; Veiko, V.P.; Kutrovskaya, S.V.; Kucherik, A.O.; Osipov, A.V.; Vartanyan, T.A.; Itina, T.E. Reliable and well-controlled synthesis of noble metal nanoparticles by continuous wave laser ablation in different liquids for deposition of thin films with variable optical properties. *J. Nanoparticle Res.* **2016**, *18*, 155. [CrossRef]

43. Takaku, R.; Wen, Q.; Cray, S.; Devlin, M.; Dicker, S.; Hanany, S.; Hasebe, T.; Iida, T.; Katayama, N.; Konishi, K.; et al. Large diameter millimeter-wave low-pass filter made of alumina with laser ablated anti-reflection coating. *Opt. Express* **2011**, *29*, 41745–41765. [CrossRef]
44. Chryssolouris, G. *Laser Machining-Theory and Practice*; Springer: New York, NY, USA, 1991; 274p.
45. Patel, R.; Chaudhary, P.S.; Soni, D.K. A review on laser engraving process for different materials. *Int. J. Sci.-Entific Res. Dev.* **2015**, *2*, 1–4.
46. Toporovskii, V.V.; Skvortsov, A.A.; Kudryashov, A.V.; Samarkin, V.V.; Sheldakova, Y.V.; Pshonkin, D.E. Flexible bi-morphic mirror with high density of control electrodes for correcting wavefront aberrations. *J. Opt. Technol.* **2019**, *86*, 32–38. [CrossRef]
47. Babu, S.; Santella, M.; Feng, Z.; Riemer, B.; Cohron, J. Empirical model of effects of pressure and temperature on electrical contact resistance of metals. *Sci. Technol. Weld. Join.* **2001**, *6*, 126–132. [CrossRef]
48. Reinert, W. Handbook of Silicon Based MEMS Materials and Technologies. *Met. Alloy. Seal Bond.* **2015**, *1*, 626–639.
49. Sprigode, T.; Gester, A.; Wagner, G.; Mäder, T.; Senf, B.; Drossel, W.-G. Mechanical and Microstructural Characterization of Ultrasonic Welded NiTiCu Shape Memory Alloy Wires to Silver-Coated Copper Ferrules. *Metals* **2021**, *11*, 1936. [CrossRef]
50. Kudryashov, A.V.; Samarkin, V.V.; Sheldakova, Y.V.; Aleksandrov, A.G. Wavefront compensation method using a Shack-Hartmann sensor as an adaptive optical element system. *Optoelectron. Instrum. Data Process.* **2012**, *48*, 153–158. [CrossRef]
51. Sheldakova, J.; Kudryashov, A.; Lylova, A.; Samarkin, V.; Byalko, A. New approach for laser beam formation by means of deformable mirrors. In *Laser Beam Shaping XVI*; SPIE: Bellingham, WA, USA, 2015; Volume 9581, p. 95810. [CrossRef]
52. Young, W.C. *Roark's Formulas for Stress and Strain*, 6th ed.; McGraw-Hill: New York, NY, USA, 1989; pp. 421–442.
53. Ellis, E.M. Low-Cost Bimorph Adaptive Mirrors. PhD Thesis, Imperial College of Science, London, UK, 1999.
54. Toporovsky, V.; Samarkin, V.; Sheldakova, J.; Rukosuev, A.; Kudryashov, A. Water-cooled stacked-actuator flexible mirror for high-power laser beam correction. *Opt. Laser Technol.* **2021**, *144*, 107427. [CrossRef]
55. Samarkin, V.; Alexandrov, A.; Galaktionov, I.; Kudryashov, A.; Nikitin, A.; Rukosuev, A.; Toporovsky, V.; Sheldakova, J. Wide-Aperture Bimorph Deformable Mirror for Beam Focusing in 4.2 PW Ti: Sa Laser. *Appl. Sci.* **2022**, *12*, 1144. [CrossRef]

Article

Development of Singular Points in a Beam Passed Phase Screen Simulating Atmospheric Turbulence and Precision of Such a Screen Approximation by Zernike Polynomials

Feodor Kanev [1,*], Nailya Makenova [1,2] and Igor Veretekhin [1]

[1] V.E. Zuev Institute of Atmospheric Optics, 634055 Tomsk, Russia; makenova@iao.ru (N.M.); aswer95@inbox.ru (I.V.)
[2] School of Energy and Power Engineering, National Research Tomsk Polytechnic University, 634050 Tomsk, Russia
* Correspondence: mna@iao.ru

Abstract: This article addresses two issues. Firstly, it was shown that if the initial phase of a Gaussian beam is specified by the sum of Zernike polynomials or by a screen simulating atmospheric turbulence, in the process of propagation, singular points appear in the wavefront of such a beam. With the use of numerical simulation, the dependence of the vortices number on the distortion characteristics and on the distance traveled by the beam was determined. The second problem analyzed in the article is the problem of a phase screen approximation by a series formed by Zernike polynomials. The carried out numerical experiments made it possible to determine the dependence of approximation accuracy on the screen parameters and on the number of polynomials entering the basis of approximation.

Keywords: adaptive optics; atmospheric turbulence; optical vortices; Zernike polynomials

Citation: Kanev, F.; Makenova, N.; Veretekhin, I. Development of Singular Points in a Beam Passed Phase Screen Simulating Atmospheric Turbulence and Precision of Such a Screen Approximation by Zernike Polynomials. *Photonics* **2022**, *9*, 285. https://doi.org/10.3390/photonics9050285

Received: 15 March 2022
Accepted: 19 April 2022
Published: 21 April 2022

Publisher's Note: MDPI stays neutral with regard to jurisdictional claims in published maps and institutional affiliations.

Copyright: © 2022 by the authors. Licensee MDPI, Basel, Switzerland. This article is an open access article distributed under the terms and conditions of the Creative Commons Attribution (CC BY) license (https://creativecommons.org/licenses/by/4.0/).

1. Introduction

The assessment of the effectiveness of correction for turbulent distortion of laser beams and the enhancement of correction quality are problems that have been studied for more than twenty years and are still actual at present [1–5]. Often, such problems are solved with the application of numerical methods, and in developed models, Zernike polynomials are often used to define the number of degrees of freedom of the active element of an adaptive system. Actually, a phase screen that defines the turbulent distortions of a beam is represented as a series of polynomials [4,5]. Therefore, it seems appropriate to carry out the numerical estimation of this expansion precision, to our knowledge, this problem has not yet been considered thoroughly.

Optical vortices (singular points or beam dislocations) are specific objects that appear sometimes in a wavefront of radiation [6]. Such objects are characterized by a small region of zero amplitude around the vortex and a cut in phase distribution. The cut ends in the center of a vortex. The development of optical vortices was observed in experiments with beams reflected from a rough surface [7,8]. At the same time, the authors of Refs [3,4] demonstrated in numerical [3] and laboratory experiments [4] that a beam propagating in a turbulent atmosphere acquires a complex form, and scintillations and intensity zeros appear in its amplitude distribution, so in the wavefront of such a beam, one can also expect the development of singular points. Beam control under such conditions presents considerable difficulties, for example, the negative influence of dislocations on the adaptive correction was theoretically demonstrated in Ref. [1].

The present article is a continuation of research in these areas. In the text, we discuss two problems. Firstly, in numerical experiments, we demonstrated that singular points appear in a wavefront of a Gaussian beam with an initial phase profile given by Zernike polynomials or by a screen simulating atmospheric turbulence, i.e., by a smooth function.

To increase reliability, the number and coordinates of dislocations were found using three detection algorithms built on different principles [9–12]. Thus, the number of vortices in the wavefront as a function of turbulence intensity and the distance traveled by the beam was obtained.

The second problem, the solution of which is discussed in the article, is the problem of phase screen approximation. It is known, that to realize successfully the adaptive control of a laser beam in a turbulent atmosphere requires the precise reconstruction of a reference beam phase. Often this phase has a complex form and includes singular points, so the reconstruction of such distribution by an adaptive mirror with a continuous surface leads to errors in the process of beam control. It is expedient to analyze this problem by numerical methods, but to do so the whole mathematical and numerical model of an adaptive system is required. This model should include many rather complex blocks, so at the thirst stage, we just considered the precision of a phase screen approximation by Zernike polynomials. The obtained results and characteristics of the method are presented in the current paper.

2. Materials and Methods

The propagation of coherent laser radiation in a turbulent medium was considered on the base of the numerical simulation technique. The optical layout of the numerical experiment is shown in Figure 1. The beam had a Gaussian amplitude profile; phase modulation was realized in the plane of the emitting aperture by a surface formed as a sum of Zernike polynomials [13] or by a phase screen simulating atmospheric turbulence. The spectral density of the index of refraction fluctuations on this screen was given by the following equation [14]:

$$\Phi_n(\kappa) = 0.033 C_n^2 (\kappa_L^2 + \kappa^2)^{-11/6} \exp(-\kappa^2/\kappa_m^2), \quad \kappa_L = 2\pi/L_0, \quad \kappa_m = 5.92/l_0. \tag{1}$$

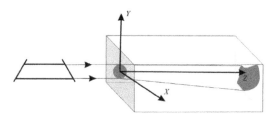

Figure 1. Layout of numerical experiments.

Equation (1) describes the von Karman spectrum of fluctuations, L_0, l_0 are the outer and inner scales of turbulence, C_n is the structure constant of atmospheric turbulence, related with Fried's coherence length by the formula $r_0 = 1.68(C_n^2 k^2 L)^{-3/5}$, here k is the wave number, L is a path length. This formula demonstrates that r_0 depends on the intensity of the index of refraction fluctuations, wavelength, and on the distance passed by a beam.

From the screen to the plane of observation the beam propagated under conditions of free diffraction. Propagation was described by the wave equation [15]:

$$2ik\left(\frac{\partial E}{\partial z} + \frac{1}{v_{gr}}\frac{\partial E}{\partial t}\right) = \Delta_\perp E, \tag{2}$$

and the fast Fourier transform was employed to solve it [16]. In Equation (2), x and y are coordinates in the plane normal to the direction of propagation; z is a coordinate along this direction; $\Delta_\perp = \partial^2/\partial x^2 + \partial^2/\partial y^2$ is the Laplace operator; v_{gr} is the group velocity of the beam. The coordinates x and y, and parameter r_0 were normalized to the beam initial radius a_0, and coordinate z to the diffraction length $z_d = ka_0^2$. All computer applications used in investigations of singular beam propagation were developed by the authors of the paper with Visual C++ language [17].

The following set of parameters was used to characterize the optical field in the plane of observation: Power-in-the-bucket (PIB).

$$J(t) = \frac{1}{P_0} \iint \rho(x,y) I(x,y,t) dxdy. \tag{3}$$

This parameter is proportional to the beam power incident in an aperture of radius S_t. In Equation (3) P_0 is the total power of the beam and

$$\rho(x,y) = \exp[-(x^2+y^2)/S_t^2]$$

is an aperture function.

The shift of the beam gravity center along axis x:

$$X_c = \frac{1}{P_0 a_0} \iint x I(x,y,t) dxdy, \tag{4}$$

and y:

$$Y_c = \frac{1}{P_0 a_0} \iint y I(x,y,t) dxdy. \tag{5}$$

The effective radius of the beam

$$R_{Eff} = \left\{ \frac{1}{P_0 a_0} \iint (\mathbf{r}_\perp - \mathbf{r}_c)^2 I(x,y,t) dxdy \right\}^{1/2}. \tag{6}$$

To obtain complete information concerning beam distortions, in addition to the above-described functions, we also registered the quantity and coordinates of optical vortices developed in the wavefront. The detailed description of the algorithms constructed to localized singular points of the wavefront was given in Ref. [9]. Algorithms were built with the use of the following special features of vortex radiation:

1. Branch cuts are present in an interference pattern of vortex radiation. On this property, the first algorithm was developed.
2. Circulation $\Gamma(\alpha)$ of wavefront gradients

$$\Gamma(\alpha) = \oint_P \alpha(\mathbf{r},z) d\mathbf{r} \tag{7}$$

is equal to $\pm 2\pi n$ if an optical vortex falls in an integration contour. The second algorithm was based on this property. Here n is a vortex topological charge and P is the perimeter of the integration contour.

3. A vortex is a point of intersections of isolines. Isolines should be drawn in distributions of real and imaginary parts of beam complex amplitude, magnitudes of corresponding functions were equal to zero along the line. On this property, the third algorithm was developed.

According to the assessments presented in Ref. [9], the highest precision of registration was achieved with algorithms 2 and 3. Computer processing of interference patterns (algorithm No. 1) does not provide high resolution, so only a small number of vortices can be localized with the application of this method.

3. Obtained Results

In the present paragraph, the results characterizing the development of singular points in a wavefront of radiation are presented. The phase profile of radiation has been formed as a sum of Zernike polynomials [14], or specified by a phase screen, simulating atmospheric turbulence. Furthermore we illustrated the quality of approximation of such a screen.

Changes in Gaussian beam amplitude are shown in Figures 2–5, while its initial phase was formed by a single polynomial. In the pictures, we can see the phase distribution of radiation (Figure 2a), cross-sections of this distribution (Figure 2b), and amplitude distribution of a beam passed a distance of 0.1 diffraction length (Figure 2c).

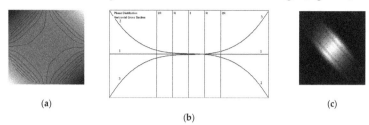

Figure 2. Phase specification by polynomial No. 5 (third-order astigmatism). Phase surface (**a**) and corresponding amplitude distribution of the beam passed path of 0.1 diffraction length (**c**) are shown by grayscale images. Cross-sections of the phase surface are presented in picture (**b**). Cross-section 1 was cut through the center of the beam, cross-section 2 corresponds to the cut shifted on the beam radius from the center in a positive direction of the y-axis, and number 3—to the cut shifted in a negative direction.

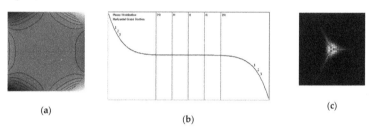

Figure 3. The same as in Figure 2, but the phase was specified by polynomial No. 9 (trefoil). Phase surface (**a**), cross-sections of phase surface (**b**) and corresponding amplitude distribution (**c**) are shown. Three cross-sections coincide (**b**).

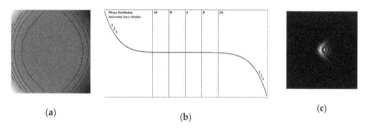

Figure 4. The same as in Figure 2, the phase was specified by polynomial No. 7 (coma). Phase surface (**a**), cross-sections of phase surface (**b**) and corresponding amplitude distribution (**c**) are shown. Three cross-sections coincide (**b**).

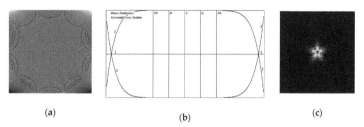

Figure 5. The same as in Figure 2, the phase was specified by polynomial No. 20. Phase surface (**a**), cross-sections of phase surface (**b**) and corresponding amplitude distribution (**c**) are shown.

In the pictures, we can see the phase distribution of radiation (Figures 2a, 3a, 4a and 5a), cross-sections of this distribution (Figures 2b, 3b, 4b and 5b), and amplitude distribution of a beam passed a distance of 0.1 diffraction length (Figures 2c, 3c, 4c and 5c).

Usually, an approximation of a screen simulating atmospheric turbulence is performed by the sum of polynomials [14], so it is feasible to consider influence induced not only by discrete components but also by the total sum of components. Corresponding examples are shown in Figures 6 and 7; in Figure 6, summation was realized from the first to sixth (trefoil) polynomials and in Figure 7 to the seventh (coma). All coefficients were the same and equal to one.

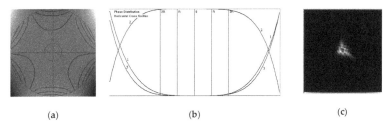

Figure 6. The same as in Figure 2, the phase was specified by the sum of polynomials from 1 to 6. Phase surface (**a**), cross-sections of phase surface (**b**) and corresponding amplitude distribution (**c**) are shown.

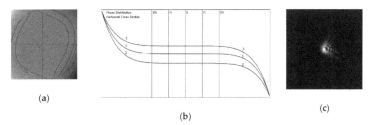

Figure 7. The same as in Figure 2, the phase was specified by the sum of polynomials from 1 to 7. Phase surface (**a**), cross-sections of phase surface (**b**) and corresponding amplitude distribution (**c**) are shown.

Registered distributions of singular points are presented in Figures 8 and 9, corresponding phase profiles were set by such polynomials as coma (Figure 8) and trefoil (Figure 9). To improve the reliability of results dislocations were localized by three algorithms built on different principles.

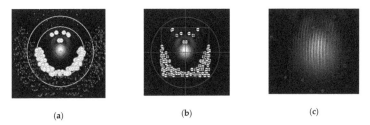

Figure 8. Distributions of singular points in a beam with a phase profile set by polynomial No. 8 (coma). The dislocations are shown by circles with «+» or «−» signs. To register optical vortices three algorithms were used. Firstly, vortices were localized by processing distribution of wavefront gradients (**a**); secondly, as a point of intersection of isolines (**b**); and with application of interferometric algorithm (**c**). The number of the found singular points N_{dsl} = 98 (**a**), N_{dsl} = 92 (**b**) N_{dsl} = 10 (**c**). The normalized path length was equal to 0.1.

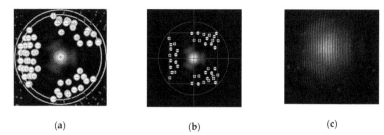

(a) (b) (c)

Figure 9. The same as in Figure 8, but the phase was specified by polynomial No. 9 (trefoil). $N_{dsl} = 58$ (**a**), $N_{dsl} = 43$ (**b**) $N_{dsl} = 7$ (**c**).

The dependence of dislocation quantity on polynomial number is shown in Figure 10. The radius of a region where detection was performed changed in the range from 1 to 1.4 initial radii of a beam.

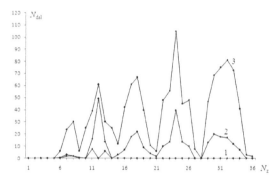

Figure 10. Number of singular points (N_{dsl}) in a beam wavefront with a phase specified by different polynomials (N_z). Registration was carried out in a region with a radius equal to the initial radius of a beam (curve 1), to two initial radii (2), to three initial radii (3). $Z = 0.1$.

The number of the found vortices as a function of the distance traveled by a beam is presented in Figures 11 and 12. The initial phase was formed by such polynomials as coma and trefoil (Figure 11) or set by a phase screen representing atmospheric turbulence (Equation (1)).

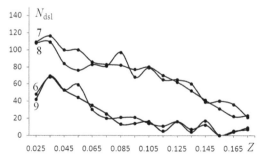

Figure 11. Dependence of dislocation number on distance Z passed by the beam. Calculation was performed for a beam with phase specified by two comas (polynomial numbers and number of curves are 7 and 8) and by trefoils (polynomials and curves are 6 and 9). The radius of registration region is 1.4 of the initial radius of the beam.

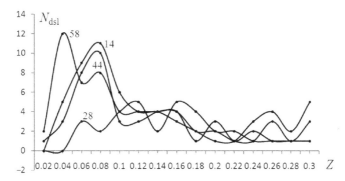

Figure 12. Dependence of dislocation number on distance Z passed by the beam. Calculation was performed for several realizations of random phase screens, numbers of which are printed near curves.

The results presented above demonstrate that the development of optical vortices is possible in a beam with an initially smooth phase profile. Let us now consider the possibility of such a profile approximation by Zernike polynomials. In the first of the two problems, the phase was represented by the sum of polynomials and approximated by the sum of polynomials, in the second, the phase was set by a turbulent screen and approximated by polynomials. In the process of approximation calculation of polynomial coefficients was realized with the least-mean-square method [15].

The first problem we divided into three variants:

The numbers of polynomials entering the basis of approximation are larger than that in the phase screen (Table 1; 12 and 9 polynomials correspondingly).

Table 1. The screen was set by 9 polynomials and approximated by 12 polynomials. Dimensions of computational grid are 256×256 nodes.

Parameters	C_{Z1}	C_{Z2}	C_{Z3}	C_{Z4}	C_{Z5}	...	C_{Z9}	J	ε_{Ph}	ε_A	R_{Eff}	X_c	Y_c
Number of the column	1	2	3	4	5	6	7	8	9	10	11	12	13
Values of parameters corresponding to the given phase	1.00	1.00	1.50	1.00	0.50	...	1.00	0.51	0.00	0.00	1.86	0.36	−0.36
Values obtained as a result of approximation	1.00	1.00	−6.11	1.00	−7.10	...	1.00	0.51	0.00	0.00	1.86	0.36	−0.36

The screen and approximation basis include the same number of functions (Tables 2–4).

Table 2. The screen was set by 9 polynomials and approximated by 9 polynomials. Dimensions of computational grid are 256×256 nodes.

Parameters	C_{Z1}	C_{Z2}	C_{Z3}	C_{Z4}	C_{Z5}	...	C_{Z9}	J	ε_{Ph}	ε_A	R_{Eff}	X_c	Y_c
Number of the column	1	2	3	4	5	6	7	8	9	10	11	12	13
Values of parameters corresponding to the given phase	1.00	1.00	1.50	1.00	0.50	...	1.00	0.51	0.00	0.00	1.86	0.36	−0.36
Values obtained as a result of approximation	1.00	1.00	−6.11	1.00	−7.11	...	1.00	0.51	0.00	0.00	1.86	0.36	−0.36

Table 3. The screen was set by 12 polynomials and approximated by 12 polynomials. Dimensions of computational grid are 256 × 256 nodes.

Parameters	C_{Z1}	C_{Z2}	C_{Z3}	C_{Z4}	C_{Z5}	...	C_{Z12}	J	ε_{Ph}	ε_A	R_{Eff}	X_c	Y_c
Number of the column	1	2	3	4	5	6	7	8	9	10	11	12	13
Values of parameters corresponding to the given phase	1.00	1.00	1.50	1.00	0.50	...	1.00	0.52	0.00	0.00	2.35	0.07	−0.07
Values obtained as a result of approximation	19.27	−3.9	−14521100		−1413.1	...	1.00	0.18	4.10	0.69	3.21	0.44	0.39

Table 4. The screen was set by 12 polynomials and approximated by 12 polynomials. Dimensions of computational grid are 2048 × 2048 nodes.

Parameters	C_{Z1}	C_{Z2}	C_{Z3}	C_{Z4}	C_{Z5}	...	C_{Z12}	J	ε_{Ph}	ε_A	R_{Eff}	X_c	Y_c
Number of the column	1	2	3	4	5	6	7	8	9	10	11	12	13
Values of parameters corresponding to the given phase	1.00	1.00	1.50	1.00	0.50	...	1.00	0.53	0.00	0.00	2.25	0.06	−0.07
Values obtained as a result of approximation	0.95	0.77	952.221.00		960.12	...	1.00	0.54	0.22	0.14	2.24	0.06	−0.07

The number of polynomials in the basis is lesser than in the phase screen (Table 5).

Table 5. The screen was set by 9 polynomials and approximated by 8 polynomials. Polynomial coefficient No. 9 was reduced to 0.2. Dimensions of computational grid are 1024 × 1024 nodes.

Parameters	C_{Z1}	C_{Z2}	C_{Z3}	C_{Z4}	C_{Z5}	...	C_{Z8}	J	ε_{Ph}	ε_A	R_{Eff}	X_c	Y_c
Number of the column	1	2	3	4	5	6	7	8	9	10	11	12	13
Values of parameters corresponding to the given phase	1.00	1.00	1.00	1.00	1.00	...	1.00	0.52	0.00	0.00	2.35	0.56	−0.41
Values obtained as a result of approximation	46.61	1.00	17.83	1.00	17.83	...	1.00	0.13	8.19	0.69	3.47	1.95	−0.36

The initial phase profile and the surface obtained as a result of approximation were formed by the same functions, so to assess the precision of calculations we can compare coefficients of the members entering the sum and deviation of one surface from another. Corresponding parameters (coefficients C_{ZN}, here, N is a polynomial number) were put in Tables 1–5. The difference between the two surfaces was calculated according to the formula

$$\varepsilon_{Ph} = \sum_{i,j}^{M} \sqrt{(A_{ij} - B_{ij})^2} / \sum_{i,j}^{M} \sqrt{(A_{ij})^2} \tag{8}$$

here, A_{ij} is the value of the given function in node i, j; B_{ij} is the value of the function obtained as a result of approximation; $M \times M$ are dimensions of the computational grid.

To increase the consistency of the solution, we compared the parameters of two beams with Gaussian amplitude distributions. The phase of the first was set by the screen formed by the sum of polynomials (Figure 1), while for the phase of the second, the surface taken obtained was in the approximation procedure. Both beams passed the same distance, after that their parameters were calculated and compared. Calculation of the following parameters was performed: power-in-the-bucket J (Equation (3)), shifts X_C and Y_C of beam gravity center along OX and OY axes (Equations (4) and (5)), the effective radius of the beam R_{Eff} (Equation (6)). Furthermore, we calculated the difference between amplitude distributions. To do so, Equation (8) was used, but instead of phase profiles, we substituted into the formula corresponding amplitude distributions.

Results obtained in the first variant are presented in Table 1. The basis of approximation was formed by 12 polynomials, and 9 polynomials were included in the phase screen.

In the second variant, an equal number of polynomials were included in the basis and screen. In Table 2, the date is presented for the screen formed by 9, and in Tables 3 and 4 by 12 polynomials.

The third variant is illustrated by the results given in Table 5 and in Figure 13.

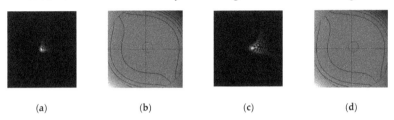

(a)　　　　　　　(b)　　　　　　　(c)　　　　　　　(d)

Figure 13. A beam amplitude registered at the distance of 0.1 diffraction length (**a**) and its initial phase (**b**) represented as the sum of the first nine polynomials. All polynomial coefficients except No. 9 were equal to one; the ninth coefficient was equal to 0.2. Amplitude (**c**) and phase (**d**) profiles obtained as a result of approximation with the use of eight polynomials.

The next problem is the approximation of the phase screen simulating turbulent fluctuations of the index of refraction. As in the previous example, propagation was analyzed by two beams with Gaussian amplitude distributions. The phase of the first was set by the phase screen, while the phase of the second was obtained as a result of approximation. Parameters of beams calculated in the plane of observations were presented in Tables 6–8; in Figures 14–16, we can see amplitude distributions and phase profiles of the beams.

Table 6. Characteristics of the beam in the plane of observations. Parameters of numeric experiment were the same as in the previous example (Figure 14).

Parameters	ε_{Ph}	ε_{Amp}	J	Sh_X	Sh_Y	R_{EffX}	R_{EffY}	R_{Eff}
Number of the column	1	2	3	4	5	6	7	8
Values of parameters corresponding to the given phase	0	0	0.988	0.043	0.019	0.053	0.084	0.099
Values obtained as a result of approximation	0.208	0.291	0.986	0.043	0.019	0.077	0.075	0.107

Table 7. Characteristics of the beam in the plane of observations calculated with increased turbulence intensity. Parameters of numeric experiment were the same as in the previous example (Figure 15).

Parameters	ε_{Ph}	ε_{Amp}	J	Sh_X	Sh_Y	R_{EffX}	R_{EffY}	R_{Eff}
Number of the column	1	2	3	4	5	6	7	8
Values of parameters corresponding to the given phase	0	0	0.983	0.076	0.035	0.038	0.092	0.100
Values obtained as a result of approximation	0.208	0.544	0.980	0.076	0.035	0.085	0.078	0.114

Table 8. Characteristics of the beam in the plane of observations. Parameters of numeric experiment were the same as in the previous example (Figure 16).

Parameters	ε_{Ph}	ε_{Amp}	J	Sh_X	Sh_Y	R_{EffX}	R_{EffY}	R_{Eff}
Number of the column	1	2	3	4	5	6	7	8
Values of parameters corresponding to	0	0	0.965	0.076	−0.149	0.039	0.081	0.090
Values obtained as a result of approximation	0.244	0.675	0.955	0.076	−0.135	0.093	0.122	0.153

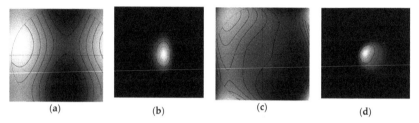

Figure 14. Precision of phase screen approximation. Phase distribution specified by the screen simulating atmospheric turbulence (**a**), amplitude distribution of a beam with initial phase profile specified by this screen (**b**), surface obtained as a result of approximation (**c**), amplitude of a beam (**d**) with initial phase presented in Figure (**c**). Dimensions of computation grid are 1024 × 1024, $r_0 = 0.2$.

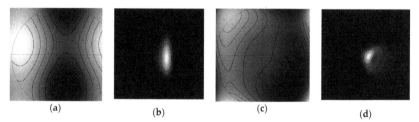

Figure 15. Increase in turbulence intensity ($r_0 = 0.1$). Other parameters of numeric experiment are the same as in Figure 14. Phase distribution specified by the screen simulating atmospheric turbulence (**a**), amplitude distribution of a beam with initial phase profile specified by this screen (**b**), surface obtained as a result of approximation (**c**), amplitude of a beam (**d**) with initial phase presented in Figure (**c**).

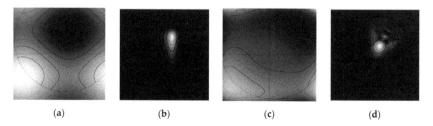

Figure 16. Results obtained with computational grid of larger (2048 × 2048 nodes) dimensions. Other parameters are the same as in the previous example (Figure 15, Table 6). Phase distribution specified by the screen simulating atmospheric turbulence (**a**), amplitude distribution of a beam with initial phase profile specified by this screen (**b**), surface obtained as a result of approximation (**c**), amplitude of a beam (**d**) with initial phase presented in Figure (**c**).

Calculations were performed with different intensities of atmospheric distortions and with the use of different calculation grids. In all numeric experiments, the inner scale of turbulence was larger than the diameter of the beam. Even under such conditions, the quality of the screen approximation was not high, but with a smaller inner scale, the obtained results were completely unsatisfying.

4. Discussion

4.1. Scintillations of Amplitude and Emergence of Singular Points in the Wavefront of a Beam with a Phase Set by Zernike Polynomials

The first block of results included in paragraph 3 of the current paper illustrates the fact that a singular wavefront can be obtained if the initial phase profile of a beam is set to be Zernike polynomials. To prove this fact, the beam with a given phase passed some distance in a non-aberrating medium, its amplitude and phase profiles were calculated in the plane of observations (the optical layout of the corresponding numerical experiment is shown in Figure 1), and singular points were registered by three different algorithms.

Figures 2–7 show that points of zero intensity appear in such a beam, the localization of optical vortices is illustrated in Figures 8 and 9. If a phase profile is formed by a sum of polynomials with equal coefficients, amplitude distribution is influenced mainly by the last polynomial entering the sum. This can be seen from the comparison of Figures 3 and 6. In the first case, the phase was set by trefoil, and in the second by the sum including polynomials from the first to trefoil. All polynomial coefficients were the same and equal to one. In both variants (Figures 3c and 6c) amplitude distributions have similar forms. The same conclusions can be drawn if we compare Figures 4c and 7c, corresponding amplitude profiles were obtained by setting the phase with coma and by sum including polynomials from the first to coma.

In the observation plane, we also survey the appearance of singular points in the wavefront of the beam. The application of two registration algorithms gives approximately the same results, i.e., close numbers of vortices and similar forms of their distribution (Figures 8a,b and 9a,b). A much smaller number of singular points were localized by the interferometric algorithm (Figures 8c and 9c) which can be explained by the low resolution by the applied technique. Nevertheless, in all situations the vortices were detected, which proves the development of singular points in a beam with an initially smooth phase profile.

Data illustrating the dependence of dislocation number on the size of the registration region, number of polynomials setting the phase profile, and the beam propagation distance are presented in Figures 10–12. As is shown in Figure 10, vortices do not appear if the phase is prescribed with the use of the first five polynomials. The singular points started to appear from the polynomial number six and higher. Comparing the form of curves in Figure 10, we can deduce that number of vortices increases with the increase in the region radius where we look for singular points.

The dependence of dislocation quantity on the distance passed by the beam is shown in Figure 11. The phase was set by polynomials. Characteristic features of all curves are oscillations and general decreases.

Approximately the same traits we observe if the phase is given by the screen simulating atmospheric turbulence (Figure 12). As in the previous graph, the curves oscillate, pass maximums, and go to zero.

4.2. Approximation of the Phase Screen Formed by Zernike Polynomials

In the previous part, we demonstrated that optical vortices appear in a wavefront if a beam passes the phase screen formed by Zernike polynomials. Let us consider how precisely we can approximate such a screen by a series of polynomials. Of course, the actual problem is the approximation of a screen simulating turbulent distortions, but to assess the precision of the method we simplified the project. The obtained results are presented in Tables 1–5. The accuracy of the method was characterized by the deviation of the given polynomial coefficients from coefficients calculated by the least-mean-square method [18].

We also compared the parameters of two beams, the phase of the first was set by the screen, and the phase of the second was obtained as the result of approximation. Both beams passed the same distance.

If the phase was set as a sum of nine polynomials and the basis of approximation was larger (Table 1, 12 polynomials in the basis) or the same (Table 2, 9 polynomials in the basis), high precision can be achieved even with small-scale (256 × 256 nodes) computational grids. In this case, values of all calculated coefficients coincided with values of given coefficients. Astigmatism is an exception, but in this case, the main influence on amplitude exerts not the magnitude but the difference between coefficients of two astigmatisms, and this difference was calculated correctly.

The precision of phase reconstruction decreases with the increase in polynomial number in the sum forming the screen (Tables 3 and 4). Unsatisfying results were obtained with the application of small-dimension grids (Table 3). A large difference was observed between values of coefficients (columns 1 and 2 of Table 3), as well as between parameters of the two beams (columns 8–13).

The accuracy of approximation can be increased with an increase in the grid dimensions (Table 4, the grid with 2048 × 2048 nodes), but notwithstanding the small difference between coefficients registered in this case (Table 4, columns 1–7) and coincidence of such integral parameters as effective radius and shift of gravity center along coordinate axes (columns 11–13), the phase of two beams differs by 22% and amplitude by 14% (columns 9 and 10).

Least-mean-square techniques also give incorrect results when the number of polynomials in the basis of approximation is smaller than in the phase screen. Corresponding data are given in Table 5 and in Figure 13. To reduce the influence of the last polynomial in the sequence forming the screen we decreased the magnitude of its coefficient from 1.0 to 0.2. However, even such lessening of its influence did not cause a coincidence in the results, the calculated parameters of the two beams were dissimilar (Table 5, columns 8–13), and the difference between their amplitude profiles can be seen by the naked eye (Figure 13).

4.3. Approximation of a Phase Screen Simulating Atmospheric Distortions

In this problem, the same model was used as in the previous part, i.e., the phase of the beam was set by a screen, then this screen is approximated by a sum of polynomials, and obtained distribution was employed as an initial phase profile of a beam with Gaussian amplitude profile. After propagation on some distance parameters, two beams were compared. Specifically, we can compare the difference in the initial phase of two beams, their amplitude distributions in the plane of observations, PIBs, shifts of gravitation centers, and so on. Corresponding parameters were put in Tables 6–8 and presented in Figures 14–16.

Comparing images in Figure 14a,c we can see that the phase profiles of two beams are not unlike, though the positions of extremums (the brightest and darkest regions) in two pictures are slightly shifted. As a result, in the plane of observations, we register the same magnitudes of PIBs (Table 6, column 3) and the same shifts of gravity centers (columns 4 and 5). The largest difference is observed in the magnitudes of effective radii (column 6–8), consequently, amplitude distributions of two beams do not also coincide (Figure 14b,c).

With the increase in turbulent intensity characterized by Fried's coherence length, the main features of the problem remain the same, the phase profiles of the two beams are similar, the effective radii and amplitude distributions are different (Table 7 and Figure 15).

The increase in the solution accuracy cannot be achieved by increasing the grid dimensions up to 2048 × 2048 nodes (Figure 16 and Table 8). In this case, the difference between amplitude distributions is approximately 68% (Table 8, column 2), effective radii along axes OX and OY of the second beam were calculated erratically (columns 6–8 of the Table), and some errors appeared in the calculation of gravity center shifts (column 5). In general, even with the large inner scale, we could not obtain the exact approximation of the phase screen simulating the influence of atmospheric turbulence. If the inner scale decreases, the quality of approximation decreases even further.

5. Conclusions

The data presented in the paper allow one to draw the following conclusions:

- If a phase profile of a beam with Gaussian distribution of amplitude is set by high Zernike polynomials, scintillations of intensity and optical vortices appear in such a beam as a result of propagation. With the increase in propagation distance, the number of vortices changes and decreases.
- Singular points of the wavefront also appear if the phase is set by a screen simulating atmospheric turbulence. Dependence on the path length is approximately the same as in the previous case: corresponding curves oscillate, reach maximums, and go to zero.

In the problem of approximation of the phase screen formed by Zernike polynomials, (approximation was also performed by the polynomials) we established that:

- If a phase screen is formed by 9 polynomials and the basis of approximation is formed by the same or greater number of polynomials high precision can be achieved even with computational grids of small dimensions (256×256 nodes).
- The quality of approximation decreases if the number of polynomials forming the screen increases (we have considered an increase in numbers up to 12). Absolutely unsatisfying results were observed on grids with small dimensions.
- The precision of the solution can be increased with the increase in the grid dimensions, but even on 2048×2048 grid the difference between given and calculated phase profiles (Equation (7)) is about 22%.

The approximation of a phase screen simulating turbulence of medium intensity showed that errors of phase restoration depend on Fried's coherence length. In all considered cases, these errors changed from 22% to 24%. Much larger (from 30% to 68%) were differences between amplitude profiles of beams.

Author Contributions: Conceptualization, software development, and original draft preparation were done by F.K. Reviewing anf editing by N.M. Calculations and analyzes of results by I.V. All authors have read and agreed to the published version of the manuscript.

Funding: Parts 2 and 3 of this research were funded by Russian Science Foundation (project No. 20-19-00597). Part 4 was funded in the framework of the State program formulated for the Institute of atmospheric optics SB RAS, Tomsk, Russia.

Institutional Review Board Statement: Npt applicable.

Informed Consent Statement: Not applicable.

Data Availability Statement: Not applicable.

Conflicts of Interest: The authors declare no conflict of interest.

References

1. Fried, D.L. Branch point problem in adaptive optics. *J. Opt. Soc. Am. A* **1998**, *15*, 2759–2767. [CrossRef]
2. Vorontsov, M.A.; Kolosov, V.V.; Kohnle, A. Adaptive laser beam projection on an extended target: Phase- and field-conjugate precompensation. *J. Opt. Soc. Am. A* **2007**, *24*, 1975–1993. [CrossRef] [PubMed]
3. Konyaev, P.A.; Lukin, V.P. Computational algorithms for simulations in atmospheric optics. *Appl. Opt.* **2016**, *55*, B107–B112. [CrossRef] [PubMed]
4. Lachinova, S.L.; Vorontsov, M.A. Laser beam projection with adaptive array of fiber collimators. II. Analysis of atmospheric compensation efficiency. *J. Opt. Soc. Am. A* **2008**, *25*, 1960–1973. [CrossRef] [PubMed]
5. Noll, R.J. Zernike polynomials and atmospheric turbulence. *J. Opt. Soc. Am.* **1976**, *66*, 207–211. [CrossRef]
6. Grier, D.G. A revolution in optical manipulation. *Nature* **2003**, *424*, 810–816. [CrossRef] [PubMed]
7. Li, X.; Tai, Y.; Zhang, L.; Li, H.; Li, L. Characterization of dynamic random process using optical vortex metrology. *Appl. Phys. B* **2014**, *116*, 901–909. [CrossRef]

8. Wang, W.; Qiao, Y.; Ishijima, R.; Yokozeki, T.; Honda, D.; Matsuda, A.; Hanson, S.G.; Takeda, M. Constellation of phase singularitie in a specklelike pattern for optical vortex metrology applied to biological kinematic analysis. *Opt. Express* **2008**, *16*, 13908–13917. [CrossRef] [PubMed]
9. Kanev, F.; Aksenov, V.P.; Veretekhin, I.D.; Makenova, N.A. Methods of optical vortex registration. In Proceedings of the 25th International Symposium on Atmospheric and Ocean Optics: Atmospheric Physics, Novosibirsk, Russia, 1–5 July 2019.
10. Patorski, K.; Pokorski, K. Examination of singular scalar light fields using wavelet processing of fork fringes. *Appl. Opt.* **2011**, *50*, 773–781. [CrossRef] [PubMed]
11. Angelsky, O.V.; Maksimyak, A.P.; Maksimyak, P.P.; Hanson, S.G. Spatial Behaviour of Singularities in Fractal-and Gaussian Speckle Fields. *Open Opt. J.* **2009**, *3*, 29–43. [CrossRef]
12. Nye, J.F. *Natural Focusing and Fine Structure of Light: Caustics and Wave Dislocations*; Institute of Physics Publishing: Bristol, PA, USA; Philadelphia, PA, USA, 1999; 328p.
13. Svechnikov, M.V.; Chkhalo, N.I.; Toropov, M.N.; Salashchenko, N.N. Resolving capacity of the circular Zernike polynomials. *Opt. Express* **2015**, *23*, 14677–14693. [CrossRef] [PubMed]
14. Chaudhary, V.; Abhilash, A. Literature review: Mitigation of atmospheric turbulence on long distance imaging system with various methods. *Int. J. Sci. Res.* **2012**, *3*, 2227–2231.
15. Andrews, L.C.; Phillips, R.L. *Laser Beam Propagation through Random Media*, 2nd ed.; SPIE Press Book: Bellingham, WA, USA, 2005; 808p.
16. Heideman, M.T.; Johnson, D.H.; Burrus, C.S. Gauss and the history of the Fast Fourier Transform. *IEEE ASSP Mag.* **1984**, *1*, 14–21. [CrossRef]
17. Lippman, S.B.; Lajoie, J.; Moo, B.E. *C++ Primer*, 5th ed.; Addison-Wesley: Boston, MA, USA, 2013; 938p.
18. Gui, G.; Adachi, F. Improved least mean square algorithm with application to adaptive sparse channel estimation. *EURASIP J. Wirel. Commun. Netw.* **2013**, *2013*, 204. [CrossRef]

Article

Local Correction of the Light Position Implemented on an FPGA Platform for a 6 Meter Telescope

Valentina Klochkova [1,2], Julia Sheldakova [2,*], Ilya Galaktionov [2], Alexander Nikitin [2], Alexis Kudryashov [2], Vadim Belousov [2] and Alexey Rukosuev [2]

[1] Special Astrophysical Observatory, Russian Academy of Sciences, 369167 Nizhnii Arkhyz, Russia; valentina.r11@ya.ru

[2] Sadovsky Institute of Geosphere Dynamics of Russian Academy of Sciences, Leninskiy Pr. 38/1, 119334 Moscow, Russia; galaktionov@activeoptics.ru (I.G.); nikitin@activeoptics.ru (A.N.); kud@activeoptics.ru (A.K.); belousov@activeoptics.ru (V.B.); alru@nightn.ru (A.R.)

* Correspondence: sheldakova@nightn.ru; Tel.: +7-9163630546

Abstract: The low-frequency component of the distortions caused by both the atmospheric turbulence and the behavior of the telescope itself has been studied. A corrector for the position of the center of the star image has been developed and is being used in front of the high-resolution Echelle spectrograph on the 6 m telescope of the Special Astrophysical Observatory, Russian Academy of Sciences. To speed up the calculations and to increase the bandwidth, a laser beam angular stabilization system based on an FPGA platform is considered. The system consists of two tip-tilt mirrors and two quadrant photodiodes. The FPGA analyzes the signals from the photodiodes, calculates and then applies the voltages to the piezo-driven tip-tilt mirrors to minimize the displacement of the beam on the photodiodes. The stabilization system was developed as a part of the adaptive optical system to improve the efficiency of the high-resolution Echelle spectrograph.

Keywords: tip-tilt mirror; FPGA; adaptive optics; angular stabilization system

Citation: Klochkova, V.; Sheldakova, J.; Galaktionov, I.; Nikitin, A.; Kudryashov, A.; Belousov, V.; Rukosuev, A. Local Correction of the Light Position Implemented on an FPGA Platform for a 6 Meter Telescope. *Photonics* **2022**, *9*, 322. https://doi.org/10.3390/photonics 9050322

Received: 15 March 2022
Accepted: 6 May 2022
Published: 8 May 2022

Publisher's Note: MDPI stays neutral with regard to jurisdictional claims in published maps and institutional affiliations.

Copyright: © 2022 by the authors. Licensee MDPI, Basel, Switzerland. This article is an open access article distributed under the terms and conditions of the Creative Commons Attribution (CC BY) license (https://creativecommons.org/licenses/by/4.0/).

1. Introduction

We consider two sources of the distortions of the light coming to the high-resolution Echelle spectrograph: the behavior of the telescope itself (its instability) and the aberrations caused by atmospheric turbulence [1]. A corrector to control for the position of the center of a star image has been developed and is being used at the entrance to the NES high-resolution spectrograph [2]. The statistics of the deviations of the position of the image center of a star have been collected for various atmospheric parameters in the surface layer (temperature, wind speed, etc.). From the practical experience of observations and statistics, it turned out that the tilts (Zernike 1 and Zernike 2) of the input beam with a diameter of 6 m are characterized by a low-frequency spectrum, with maximum deviations in the position of a star depending on the image quality.

Quasi-periodic fluctuations of the position of a star on the spectrograph slit were detected during the first observations at the Nasmyth foci of the 6 m telescope. At the very beginning, we believed that those fluctuations were caused by the restless behavior of the astronomer, who was on the moving platform of the Nasmyth focus, but this was proved to be false. It was also found that the excessively frequent correction of the position of the star through the telescope's driving system could lead to slowly damped oscillations or even rock the telescope. The relationship between the nature of fluctuations and the speed and duration of a single act of correction was also revealed. The effect of slowly damped oscillations on the position of a star was also discovered as a result of short-term wind gusts (up to a speed of several m/s). At the minimum speed of the correction of the movement of the telescope, oscillations occurred, with an amplitude of up to 1″, and the damping time of the oscillations was 25 s. As the correction speed increased, the effect became stronger. It was concluded that, to keep the center of the star image at the center of the spectrograph slit, it was optimal to use a local corrector, which would eliminate the need to correct the overall tilt of the telescope at high (for a telescope) frequencies.

It is already an established practice in astronomy to use an image-stabilization system [3–8]. For example, a fast-steering mirror was produced and designed to correct for the image stabilization of the astronomical telescopes [3]; this work describes the optimization process of mirror development and the system performance. Another approach, when one piezo-driven tip-tilt mirror was used to minimize the motions of the solar optical telescope images with a frequency above 14 Hz to improve polarimetric measurements, is described in [4]. A tip-tilt controller for the Solar Orbiter is discussed in [5]. A control algorithm was developed by the authors as well [6]. An image-stabilization system was described recently [7] by Chinese scientists. They presented both simulations and experiments on testing the accuracy of the image-stabilization system of a space telescope. The influence of the vibration isolation platform was also considered. Moreover, there are a number of works that are not related directly to astronomy but are of particular interest for beam stabilization [9,10].

This paper discusses a system for active laser beam stabilization based on an FPGA [11]. The system is a part of a closed-loop adaptive optical system to improve the efficiency of high-resolution spectroscopy on the 6 m telescope BTA [12]. The structure of the overall adaptive system for the telescope is presented in Figure 1. Each part is used for a certain purpose: some mirrors serve to compensate for large-scale slowly changing aberrations, while other mirrors compensate for small-scale aberrations that occur when passing through a turbulent atmosphere, and some mirrors are used to stabilize the position of the beam. We expect the whole adaptive system to decrease a star image on the spectrograph slit from 1 arcsec to 0.2 arcsec.

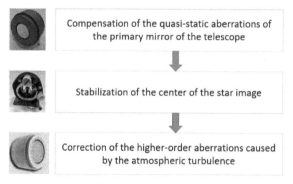

Figure 1. A cascade of adaptive systems to increase the efficiency of high-resolution spectroscopy.

At the initial stage of our work, we analyzed the structure of the aberrations in terms of Zernike polynomials and the typical amplitudes of Zernike coefficients measured in our laboratory setup [13]. Experiments showed that the contribution of the tilts is essential, so the beam position should be separately controlled during the wavefront correction of a beam distorted by the atmospheric turbulence. The experimental setup consisted of two tip-tilt mirrors, two control units to apply voltages to the piezo drivers, and two quadrant photodiodes to detect the beam position jitter. The first mirror is needed to set the beam to the desired point, and the second one facilitates directing the beam along the optical axis. The use of the FPGA allows the combination of all the electronics in one box, and the main goal of the FPGA is to reduce calculation time so that the tip-tilt correctors become the main limiting element of the system. The FPGA collects data from the quadrant photodiodes, analyzes the data, and calculates and applies voltages to the tip-tilt mirror's drivers. A scheme of a digital control module and its interaction with the tip-tilt mirrors and control units is elaborated. To confirm the ability of the beam stabilization, a series of experiments were carried out.

2. Atmospheric Turbulence Input

To analyze the input of the atmospheric turbulence, we used a setup based on a self-made Shack–Hartmann wavefront sensor [14–18]. A laser beam wavefront is distorted by the airflow made by a 1.5 kW fan heater with a diameter of 200 mm. The aberrations caused by the fan heater are then analyzed with a high-speed Shack–Hartmann wavefront sensor (Figure 2) [19]. The image resolution is 480 × 480, and the microlenses array has 20 × 20 lenses with a focal length of 12 mm and a size of 240 µ × 240 µ each. A laser diode (LDS-655-FP-1.5 by Laserscom) with a wavelength of 650 nm coupled to an optical fiber is collimated up to a diameter of 50 mm. The data is obtained with a frequency of 2000 Hz. To reconstruct the wavefront, a modal method is used [20,21]. More information on the experiment is presented in [13]. The spectral energy distribution over Zernike polynomials is presented in Figure 3. This picture confirms that not only the tilts have to be corrected but also that higher-order aberrations have sufficient input. Thus, the use of the tip-tilt correction system is only the first step, and higher-order adaptive optics should be used [22] when the tilts are compensated. The tilts (Zernike 1 and Zernike 2) have the largest amplitude among the distortions caused by the atmospheric turbulence [23].

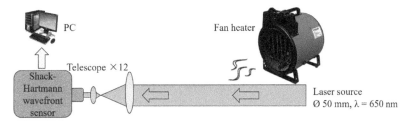

Figure 2. Experimental setup for wavefront statistical characteristics investigation.

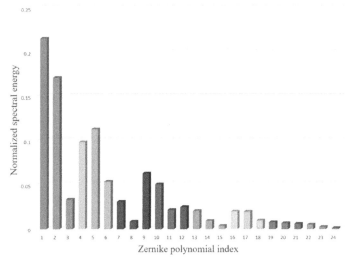

Figure 3. Distribution of turbulence energy over Zernike polynomials 1–24.

The displacements of the center of the star image in front of the spectrograph of the 6 m telescope were also measured, and results are presented in Figure 4. The fluctuations are caused not only by the turbulence but also by the fundamental frequency of the telescope, the non-ideal position of the axes, mechanical deformations, and the wind inside the telescope dome. The center of the image fluctuates with a typical amplitude of ±1 arcsec. As the spectrograph slit is just about 0.7 arcsec, these fluctuations of the image lead to 80% loss of the incoming light. If the light is stabilized with an accuracy of ±0.1–0.2 arcsec, it allows an increase in the spectrograph transmission up to 2.5 times, which is equivalent to one order of magnitude gain in angular resolution.

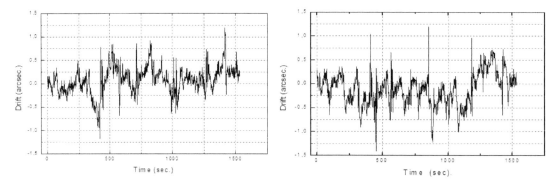

Figure 4. Drift of the star image center along X axis (**left**) and Y axis (**right**).

When the weather conditions are bad and the wind is strong, the amplitude of the fluctuations of the position of a star image increases (Figure 5). Here we can also visualize the total drift of the star image over tens of seconds.

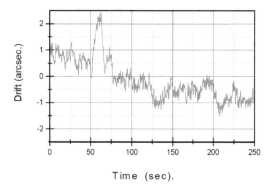

Figure 5. Drift of the star image center along X on a windy night.

3. FPGA-Based Light Position Stabilization
3.1. Tip-Tilt Mirror Structure

To stabilize the beam position, two tip-tilt mirrors controlled with an FPGA board were used. The tilts of the mirrors were ensured by piezoceramic elements with a capacity of 1 mF. The diameter of the mirror was 60 mm. The first resonance frequency of the mirror was found at 615 Hz. The mirror could be controlled with a frequency up to 500 Hz. The full dynamic range for the tilt correction was equal to 200 μrad, with a resolution of 0.05 μrad. A photo of the mirror and a cross-section together with the back view are shown in Figure 6. The piezoceramic elements (Ring Multilayer Piezo Actuators PTH1500525101 by Suzhou Pant Piezoelectric Tech. Co) were combined with the adjustment screws (9SM127M-10 by Standa).

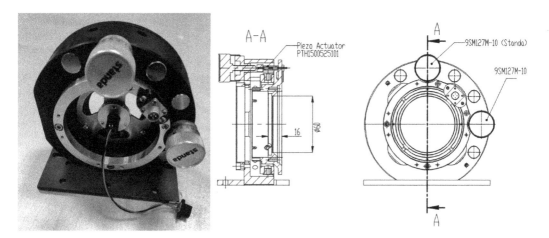

Figure 6. Photo and drawings of the tip-tilt mirror.

3.2. Beam Stabilization Setup

The setup to investigate the efficiency of the work of the beam-stabilization system is presented in Figure 7. The 50 mm laser beam distorted by a fan heater reflected from two tip-tilt mirrors, and the center of the beam was then detected with two quadrant photodiodes with a diameter of 10.2 mm, each placed at different distances from the light source (25 cm in our case). The data from the photodiodes was collected by the FPGA, and then the voltages to be applied to the piezo drivers of the mirrors to correct for the beam displacement were calculated. The FPGA sent commands to the control units (CU 1, CU 2) to change the voltages on the piezo drivers of the mirrors. Control units allow the application of voltages from -30 V to 190 V. We used a PC to adjust the system before the correction was started, to measure the response functions of each tip-tilt mirror, and to save the resulting data in an appropriate format.

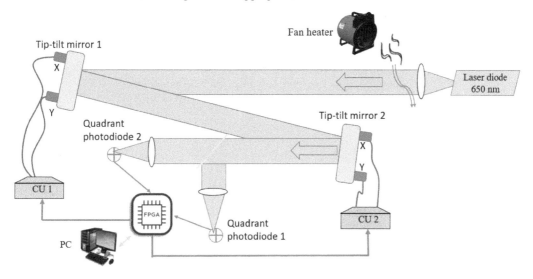

Figure 7. Beam-stabilization system.

3.3. FPGA Implementation

The FPGA was placed on a printed board to make one digital module connected with the external modules to digitize the voltages from the photodiodes, to calculate the voltages for the tip-tilt drivers, to make a digital-to-analog conversion of the voltages, and also to synchronize the entire system. The board consisted of the following parts (Figure 8):

- An analog-to-digital conversion unit to receive the voltages from the quadrant photodiodes;
- Unit for calculating digital values of corrective voltages;
- Unit for digital-to-analog conversion.

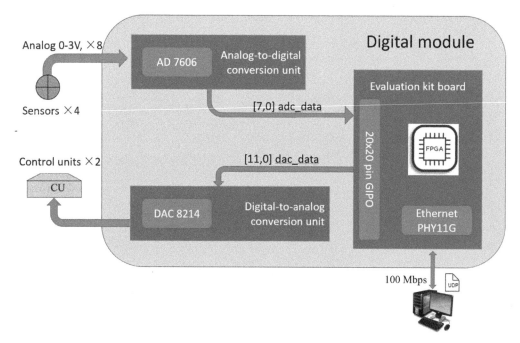

Figure 8. Scheme of the digital module.

Each unit was placed on its own PC board. The elements were connected by a digital channel with the LVCMOS 2.5 V interface. The Cyclone 10 LP FPGA Evolution Kit board FPGA Intel® Cyclone® 10CL025YU256 was used as a unit to calculate the value of the voltages for piezo drivers.

The FPGA configuration was developed in the Quartus Prime 18.0 CAD environment from Intel. The configuration consists of 4 main modules:

- Control and communication module with an analog-to-digital conversion (ADC) unit
- Control and communication module with a digital-to-analog conversion (DAC) unit
- Module for receiving and sending data to the control program installed on the PC via the UDP protocol
- Control module for the operation of the block for calculating digital values of the corrective voltages

The control and communication module with an analog-to-digital conversion unit had eight channels for converting an analog signal into a digital code. It also facilitated working with 1-bit, 8-bit, and 16-bit data buses. In the described system, due to the limited size of the connector, the mode with an 8-bit data bus was used. The module controlled the operation of the chip, receiving and buffering data.

The control and communication module with the digital-to-analog conversion unit provided sequential loading of the values into each channel of the DAC chip in double-buffering mode for the simultaneous setting of the voltages in all analog channels of the chip.

The communication module contained internal registers to interact with the control module for calculating digital values of the voltages. The operation of the module was independent of the main operation cycle of the tip-tilt system.

The control module analyzed the state of the control registers of the communication module with the control program. Upon detection of the value corresponding to the start of the work cycle, the module performed the following actions:

- Sent a start command to the control and communication module with the analog-to-digital converter;
- Wrote registers with the data, on command from the analog-to-digital conversion unit;
- Calculated voltages values;
- Sent a command to the control and communication module with a digital-to-analog converter, together with the calculated voltages;
- Updated the values of the registers of the communication module with the control program;
- Started the pause counter;
- Repeated actions when the value of the pause counter, set by the control program, was reached in the corresponding register.

The cyclogram of the control module is presented in Figure 9. The overall time for the implementation of the main operations took 62.1 µs; the most time-consuming operation (60 µs) was receiving data from the analog-to-digital converter. The limit frequency of such a system was 15 kHz; however, when changing the analog-to-digital converter operating mode to 16-bit, 30 kHz frequency could be reached. "Pause" on the cyclogram related to the tip-tilt mirror properties and was not taken into account; only control module characteristics were estimated. Based on the amplitude–resonance characteristics of the mirror, the "pause" should be about 2 ms to avoid mirror oscillations, which led to the frequency of the overall system to be 500 Hz.

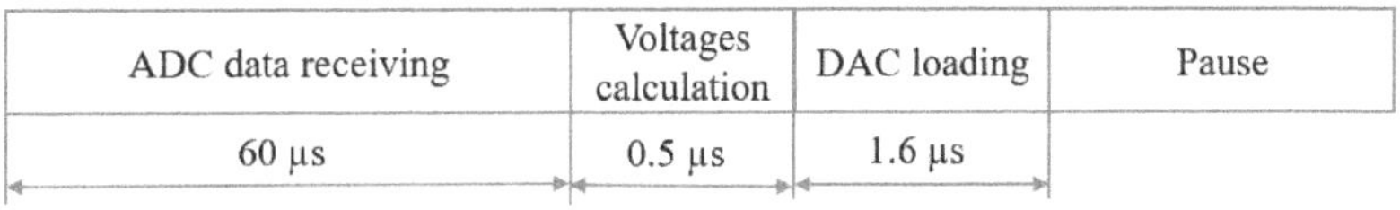

ADC data receiving	Voltages calculation	DAC loading	Pause
60 µs	0.5 µs	1.6 µs	

Figure 9. Cyclogram of the control module.

3.4. Correction Algorithm

To stabilize the laser beam in a certain place and direction, the control module should continuously determine the deviation of the laser beam from it and drive the tip-tilt mirrors' actuators. For example, a paraboloid interpolation algorithm was used in [6] to estimate an image jitter at high resolution. A Jacobian matrix feedback controller was proposed in [10], which was adaptive to the changes of the beam path. A centroid algorithm generated inverse response to the tip-tilt aberration to control for the tip-tilt mirror at UC Mount John Observatory [24].

To calculate the control voltages, we used an algorithm based on the least square method [25]. The information on the beam position was collected from the quadrant photodiodes (Figure 10). We used FD-20KP quadrant photodiodes with a diameter of 10.2 mm and a gap of 0.3 mm between the four sectors. A standard 4-channel current-to-voltage conversion circuit on an operational amplifier [26] with a conversion factor of 1,000,000 was produced to transmit the signals from each quadrant photodiode. The beam diameter on the photodiodes was 5 mm.

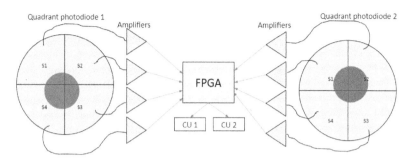

Figure 10. Beam position detection.

First, the equations set was composed as follows:

$$X1 = a11 \times U1 + a12 \times U2 + a13 \times U3 + a14 \times U4, \tag{1}$$

$$Y1 = a21 \times U1 + a22 \times U2 + a23 \times U3 + a24 \times U4, \tag{2}$$

$$X2 = a31 \times U1 + a32 \times U2 + a33 \times U3 + a34 \times U4, \tag{3}$$

$$Y2 = a41 \times U1 + a42 \times U2 + a43 \times U3 + a44 \times U4, \tag{4}$$

where U1–U4 are voltages to be applied to the drivers of the mirrors, and coefficients a11–a44 are normalizing values proportional to the response of the actuators, and X1/X2 and Y1/Y2 are the positions of the beam center on the photodiodes calculated as

$$X = ((S2 + S3) - (S1 + S4))/(S1 + S2 + S3 + S4), \tag{5}$$

$$Y = ((S1 + S2) - (S3 + S4))/(S1 + S2 + S3 + S4), \tag{6}$$

S1–S4 are levels of the signal on the segments of the quadrant photodiodes.

The response functions of the actuators were measured when a unit voltage, for example, U1 = 10 V, was applied to one actuator of the mirror. When U1 was applied to the actuator #1 (X-actuator of the first tip-tilt mirror), the displacements of the focal spots on the photodiodes ΔX1(U1), ΔX2(U1), ΔY1(U1), and ΔY2(U1) were measured, and coefficients a11–a14 could be calculated as

$$a11 = \Delta X1/U1,\ a21 = \Delta Y1/U1,\ a31 = \Delta X2/U1,\ a41 = \Delta Y2/U1, \tag{7}$$

where ΔX1/ΔX2 and ΔY1/ΔY2 are the displacements of the focal spot on the photodiodes. The same procedure was repeated for each of the four actuators, and, as a result, matrix A was filled as

$$A = \begin{bmatrix} a11 & \cdots & a14 \\ \vdots & \ddots & \vdots \\ a41 & \cdots & a44) \end{bmatrix}. \tag{8}$$

With response functions measurements, the PC sent commands to the FPGA to set the voltages and to read the coordinates of the beam center. Then an inverse matrix $|B| = |A|^{-1}$ was calculated by the PC, and the PC sent this matrix to the FPGA. The system was then calibrated.

The correction cycle includes three steps performed by the FPGA only:

1. The FPGA reads the signals from the photodiodes and determines the coordinates of the beam center on two photodiodes of X1, X2, Y1 or Y2, using Equations (5) and (6).
2. The voltages are calculated as

$$|U| = |B| \times |\ X1\ Y1\ X2\ Y2\ | \tag{9}$$

3. The voltages are applied to the actuators of the tip-tilt correctors.

4. Results and Discussion

A photo of the setup to test the abilities of the stabilizing system is presented in Figure 11. The laser beam is reflected from the tip-tilt mirrors and then detected with the quadrant photodiodes. The signal from the photodiodes is transferred to the FPGA. The FPGA analyzes the data and calculates and sends voltages to the drivers of the tip-tilt mirrors. The displacement of the beam position is saved for further analysis.

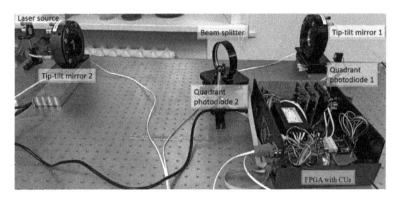

Figure 11. Photo of the testbed.

Figure 12 shows the measured fluctuation of the displacements on the first photodiode for 500 ms; the red line corresponds to the beam center position without any correction, and the green line shows the behavior of the beam center when correction is on.

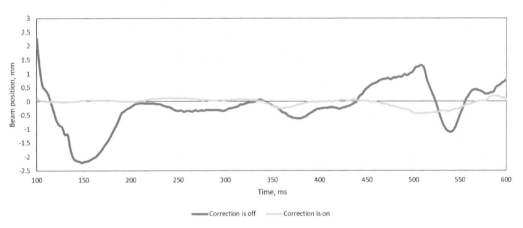

Figure 12. Fluctuations of the beam position on the 1st quadrant photodiode.

The accuracy of the center position calculation on each photodiode was 0.25 μm. As the distance between the two photodiodes was 25 cm, the accuracy of the beam positioning was ±2 μrad.

The RMS of the beam position was calculated during 30 s both with and without correction. The RMS without correction was 1 mm, and it was 0.3 mm with the correction loop closed. The ratio between these data was $20 \lg(RMS_{off}/RMS_{on})$, which showed that the system allowed the reduction of the amplitude of the beam position change up to 10 dB.

Regarding the electronics and mechanical elements, it is possible to achieve a required jitter reduction better than 10 dB. To improve the result, the beam diameter on the photodiodes should be optimized to provide a linear response at the positions close to the photodiode center; additionally, the type of photodiodes should be chosen, taking into account their signal-to-noise characteristics. One of the weak points might be an algorithm of correction based on proportional terms only. As a next step, it is necessary to modify the algorithm and integral and derivative terms.

For sure our laboratory setup was not equal to the real conditions at the telescope—herein, we wanted to present our results on a comparatively low-cost, reliable, and simple beam-stabilization system that could be installed on a 6 m telescope.

5. Conclusions

The FPGA platform was used to implement a beam-position-stabilization system. The system consists of two tip-tilt piezo-driven mirrors with control units and two quadrant photodiodes. The signals from the photodiodes are collected with the FPGA, and the control voltages necessary for the correction of the beam center displacement are calculated and applied to the actuators of the mirrors. The FPGA allows the speeding up of data processing and DAC/ADC data conversion, and the system performance only depends on the properties of the tip-tilt corrector and could be improved if necessary. The proposed beam-stabilization system allows reducing the beam jitter up to 10 dB. This number is actually not the expected one (for the BTA telescope, we need at least 5 times the jitter reduction), but we have the ideas and a way to improve it.

Author Contributions: Conceptualization, V.K. and J.S.; methodology, A.R.; software, V.B.; validation, I.G., A.N. and A.R.; formal analysis, A.R.; investigation, A.N.; resources, A.R.; data curation, A.K.; writing—original draft preparation, I.G.; writing—review and editing, A.K.; visualization, I.G.; supervision, V.K.; project administration, J.S.; funding acquisition, J.S. and A.K. All authors have read and agreed to the published version of the manuscript.

Funding: The investigations of the local correction or the beam position (Sections 3 and 4) were supported by the Russian Science Foundation under grant #20-19-00597, and the artificial turbulence investigation (Section 2) was supported by the state assignment of the Ministry of Science and Higher Education of the Russian Federation (theme No. 1021052706254-7-1.5.4).

Institutional Review Board Statement: Not applicable.

Informed Consent Statement: Not applicable.

Data Availability Statement: Not applicable.

Conflicts of Interest: The authors declare no conflict of interest.

References

1. Ivanov, A.A.; Panchuk, V.E.; Shergin, V.S. *SAO Preprint No. 155*; Spec. Astrophys. Observ: Nizhny Arkhyz, Russia, 2001; pp. 1–19.
2. Panchuk, V.; Klochkova, V.; Yushkin, M. The high-resolution echelle spectrograph of the 6-m telescope of the special astrophysical observatory. *Astron. Rep.* **2017**, *61*, 820–831. [CrossRef]
3. Dong, Z.; Jiang, A.; Dai, Y.; Xue, J. Space-qualified fast steering mirror for an image stabilization system of space astronomical telescopes. *Appl. Opt.* **2018**, *57*, 9307. [CrossRef] [PubMed]
4. Shimizu, T.; Nagata, S.; Tsuneta, S.; Tarbell, T.; Edwards, C.; Shine, R.; Hoffmann, C.; Thomas, E.; Sour, S.; Rehse, R.; et al. Image stabilization system for Hinode (Solar-B) solar optical telescope. *Sol. Phys.* **2008**, *249*, 221–232. [CrossRef]
5. Casas, A.; Gómez Cama, J.M.; Roma, D.; Carmona, M.; Bosch, J.; Herms, A.; Sabater, J.; Volkmer, R.; Heidecke, F.; Maue, T.; et al. Design and test of a tip-tilt controller for an image stabilization system. *Proc. SPIE* **2016**, *9911*, 991123. [CrossRef]
6. Roma, D.; Carmona, M.; Bosch, J.; Casas, A.; Herms, A.; Lopez, M.; Ruiz, O.; Sabater, J.; Berkefeld, T.; Maue, T.; et al. Subpixel real-time jitter detection algorithm and implementation forpolarimetric and helioseismic imager. *J. Astron. Telesc. Instrum. Syst.* **2019**, *5*, 039003. [CrossRef]
7. Chenghao, L.; Xu, H.; Qi, J.; Xiaohui, Z.; Kuo, F. Theoretical and experimental study on the testing accuracy of the image stabilization system of a space astronomical telescope. *Appl. Opt.* **2020**, *59*, 6658–6670. [CrossRef]
8. Canuto, E.; Musso, F. Active angular stabilization of the GAIA space telescope through laser interferometry. *IFAC Proc. Vol.* **2004**, *37*, 955–960. [CrossRef]

9. Tyszka, K.; Dobosz, M. Laser beam angular stabilization system based on a compact interferometer and a precise double-wedge deflector. *Rev. Sci. Instrum.* **2018**, *89*, 085121. [CrossRef]
10. Chang, H.; Ge, W.-Q.; Wang, H.-C.; Yuan, H.; Fan, Z.-W. Laser beam pointing stabilization control through disturbance classification. *Sensors* **2021**, *21*, 1946. [CrossRef]
11. Sheldakova, J.; Galaktionov, I.; Nikitin, A.; Alexandrov, A.; Kudryashov, A.; Belousov, V.; Rukosuev, A. FPGA based laser beam stabilization system. *Proc. SPIE* **2022**, *11987*, 119870C. [CrossRef]
12. Klochkova, V.G.; Sheldakova, Y.V.; Vlasyuk, V.V.; Kudryashov, A.V. Improving the efficiency of high-resolution spectroscopy on the 6-m telescope using adaptive optics techniques. *Astrophys. Bull.* **2020**, *75*, 468–481. [CrossRef]
13. Rukosuev, A.; Nikitin, A.; Belousov, V.; Sheldakova, J.; Toporovsky, V.; Kudryashov, A. Expansion of the laser beam wavefront in terms of zernike polynomials in the problem of turbulence testing. *Appl. Sci.* **2021**, *11*, 12112. [CrossRef]
14. Neal, D.R.; Pulaski, P.; Raymond, T.D.; Neal, D.A. Testing highly aberrated large optics with a Shack-Hartmann wavefront sensor. *Proc. SPIE* **2003**, *5162*, 129–138. [CrossRef]
15. Mansuripur, M. The Shack-Hartmann Wavefront sensor. *Opt. Photonics News* **1999**, *10*, 48–51. [CrossRef]
16. Wilson, R.W. SLODAR: Measuring optical turbulence altitude with a Shack–Hartmann wavefront sensor. *Mon. Not. R. Astron. Soc.* **2002**, *337*, 103–108. [CrossRef]
17. Primot, J. Theoretical description of Shack–Hartmann wave-front sensor. *Opt. Commun.* **2003**, *222*, 81–92. [CrossRef]
18. Nikitin, A.; Sheldakova, J.; Kudryashov, A.; Borsoni, G.; Denisov, D.; Karasik, V.; Sakharov, A. A device based on the Shack-Hartmann wave front sensor for testing wide aperture optics. *Proc. SPIE* **2016**, *9754*, 97540K. [CrossRef]
19. Kudryashov, A.; Rukosuev, A.; Nikitin, A.; Galaktionov, I.; Sheldakova, J. Real-time 1.5 kHz adaptive optical system to correct for atmospheric turbulence. *Opt. Express* **2020**, *28*, 37546–37552. [CrossRef]
20. Southwell, W.H. Wave-front estimation from wave-front slope measurements. *J. Opt. Soc. Am.* **1980**, *70*, 998–1006. [CrossRef]
21. Neal, D.R.; Copland, J.; Neal, D.A. Shack-Hartmann wavefront sensor precision and accuracy. *Proc. SPIE* **2002**, *4779*, 148–160. [CrossRef]
22. Hippler, S. Adaptive optics for extremely large telescopes. *J. Astron. Instrum.* **2019**, *8*, 1950001. [CrossRef]
23. Noll, R.J. Zernike polynomials and atmospheric turbulence. *J. Opt. Soc. Am.* **1976**, *66*, 207–211. [CrossRef]
24. Liu, J.; Muruganandan, V.; Clare, R.; Ramirez Trujillo, M.C.; Weddell, S. A tip-tilt mirror control system for partial image correction at UC mount john observatory. In Proceedings of the 35th International Conference on Image and Vision Computing New Zealand (IVCNZ), Wellington, New Zealand, 25–27 November 2020; pp. 1–6. [CrossRef]
25. Kudryashov, A.; Alexandrov, A.; Rukosuev, A.; Samarkin, V.; Galarneau, P.; Turbide, S.; Châteauneuf, F. Extremely high-power CO_2 laser beam correction. *Appl. Opt.* **2015**, *54*, 4352–4358. [CrossRef] [PubMed]
26. Horowitz, P.; Hill, W. *The Art of Electronics*, 3rd ed.; Cambridge University Press: New York, NY, USA, 2015.

MDPI
St. Alban-Anlage 66
4052 Basel
Switzerland
Tel. +41 61 683 77 34
Fax +41 61 302 89 18
www.mdpi.com

Photonics Editorial Office
E-mail: photonics@mdpi.com
www.mdpi.com/journal/photonics

CPSIA information can be obtained
at www.ICGtesting.com
Printed in the USA
BVHW011228270123
657290BV00008B/827